MW00849851

CORVETTE STINGRAY

THE MID-ENGINE EVOLUTION

FOREWORD BY MARK REUSS

motorbooks

CORVETTE STINGRAY

THE MID-ENGINE EVOLUTION

Quarto.com

© 2020, 2024 Quarto Publishing Group USA Inc.
Text © 2020, 2024 General Motors & Richard Prince

First Published in 2020. Second edition publushed by Motorbooks, an imprint of The Quarto Group,
100 Cummings Center, Suite 265-D, Beverly, MA 01915, USA.
T (978) 282-9590 F (978) 283-2742

All rights reserved. No part of this book may be reproduced in any form without written permission of the copy-right owners. All images in this book have been reproduced with the knowledge and prior consent of the artists concerned, and no responsibility is accepted by producer, publisher, or printer for any infringement of copyright or otherwise, arising from the contents of this publication. Every effort has been made to ensure that credits accurately comply with information supplied. We apologize for any inaccuracies that may have occurred and will resolve inaccurate or missing information in a subsequent reprinting of the book.

Motorbooks titles are also available at discount for retail, wholesale, promotional, and bulk purchase. For de-tails, contact the Special Sales Manager by email at specialsales@quarto.com or by mail at The Quarto Group, Attn: Special Sales Manager, 100 Cummings Center, Suite 265-D, Beverly, MA 01915, USA.

28 27 26 25 24   1 2 3 4 5

ISBN: 978-0-7603-8828-0

Digital edition published in 2024
eISBN: 978-0-7603-8829-7

Library of Congress Cataloging-in-Publication Data available

Design:
Cover Image:
Page Layout:
Photography:

General Motors trademarks used under license to Quarto Publishing Group USA Inc.

Printed in China

women and men at General Motors who devote themselves to making Corvette the best sports car in the world, and to the legions of enthusiasts around the globe who love America's sports car as much as I do.

CONTENTS

FOREWORD
BY MARK REUSS

You hold in your hands the story of the Corvette that changed everything. The 2020 Chevrolet Corvette Stingray is, true to the vision of creator and original Chief Engineer Zora Arkus-Duntov, a mid-engine supercar. The eighth-generation Corvette, or C8, is everything Zora dreamed of . . . with technology he never could have imagined.

General Motors has explored mid-engine concepts dating back to the original Chevrolet Engineering Research Vehicle (CERV I) from 1960. Zora famously advocated for mid-engine vehicles, but we needed to make sure we kept Corvette true to its roots of attainable performance. Mid-engine has historically posed a challenge to this mission. Not so, anymore. The time has come, today, and we feel certain that both Corvette traditionalists and potential new customers will embrace the change in layout.

The reason for that confidence becomes clear as soon as you drive it. You will think it's flat-out the best Corvette you have ever driven. And that's because it's the best Corvette anybody's ever driven. There are many reasons for that, even beyond the mid-engine layout . . . reasons like the way it feels, the way it sounds, the way it looks and the incredible attention to every detail.

Since 1953, through the good times and the bad for this company, there was always Corvette, demonstrating what it means to win.

And with every succeeding generation since 1953, Chevrolet has worked to make Corvette better and better. We have never stopped improving, never stopped innovating and never stopped making the car faster, better handling, more comfortable—more everything.

Once we got to C7, we had pushed the limits of what we could do with that configuration. It was as close to perfection as a front-engine/rear-drive Corvette was going to get. To take performance and driving dynamics to the next level for our customers, we had to move to mid-engine. As a result, the C8, to me, with its perfect balance, superior handling and outstanding design, is the best Corvette ever.

It's not a statement I make lightly . . . Corvette is the reason I work at General Motors, having grown up in one. On many weekend afternoons, my dad took me to work with him at the GM Technical Center in Warren, Michigan, specifically the Chevrolet Engineering Building, where the Corvette came to life. I spent many of those rides to Warren hunched in the rear flat area of a Corvette coupe.

This is personal for me—as it is for Corvette enthusiasts the world over. The passion all of us feel for this vehicle is the main ingredient in the recipe we used to create the C8. This book deftly tells the story of exactly how it all happened. I'd like to thank the writers, editors and artists who brought the C8 story to life so vividly, and I'd like to thank Corvette fans everywhere who love and support the vehicle so passionately. This book is for you.

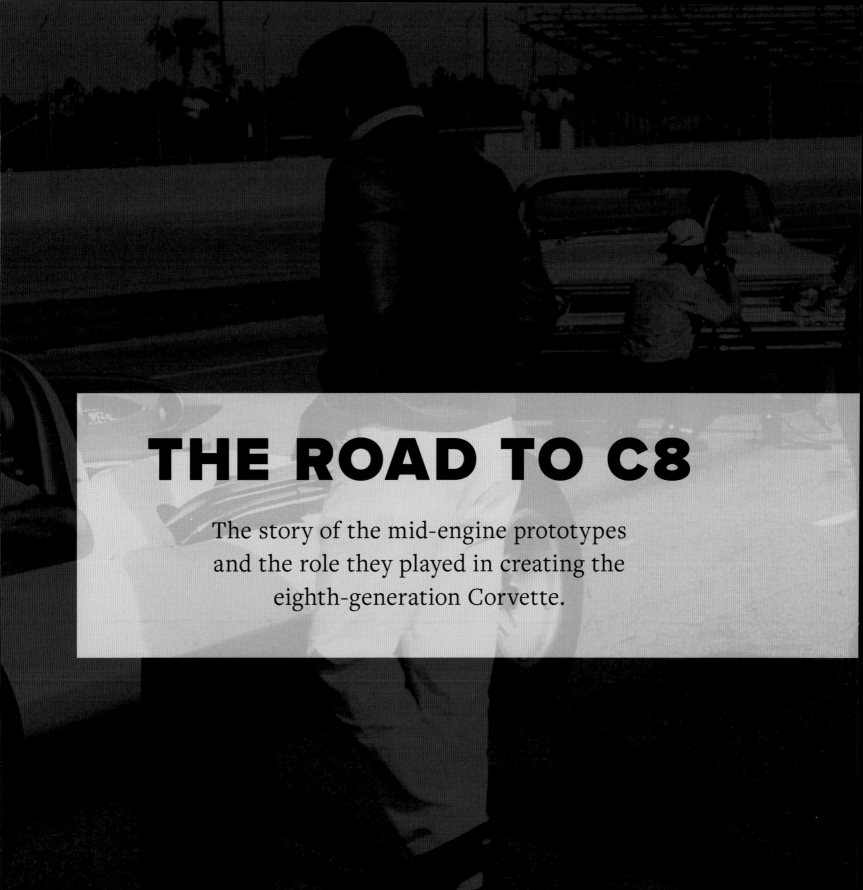

THE ROAD TO C8

The story of the mid-engine prototypes
and the role they played in creating the
eighth-generation Corvette.

The road to C8 takes us

back more than 60 years, to the late 1950s. Even though Chevrolet had pulled out of factory racing programs in 1957, Zora Arkus-Duntov and the Chevrolet engineering team still wanted to build the most daring Corvettes they could and didn't hesitate to use racing technology to make it possible.

They also wanted to prove to the world GM still had the right stuff to go racing if it so chose. This resulted in a series of prototypes. Some were engineering-based like CERV I (Chevrolet Engineering Research Vehicle) while others were design exercises meant to test public reaction to a potential new production Corvette down the road. These cars created a speculative frenzy in the press and often appeared on magazine covers.

CLOCKWISE FROM TOP LEFT: An early scale model of CERV I; John Fitch, left, with Zora-Arkus Duntov; the Aerovette concept car.

NEXT SPREAD: Arkus-Duntov drives the CERV I up Pikes Peak in Colorado.

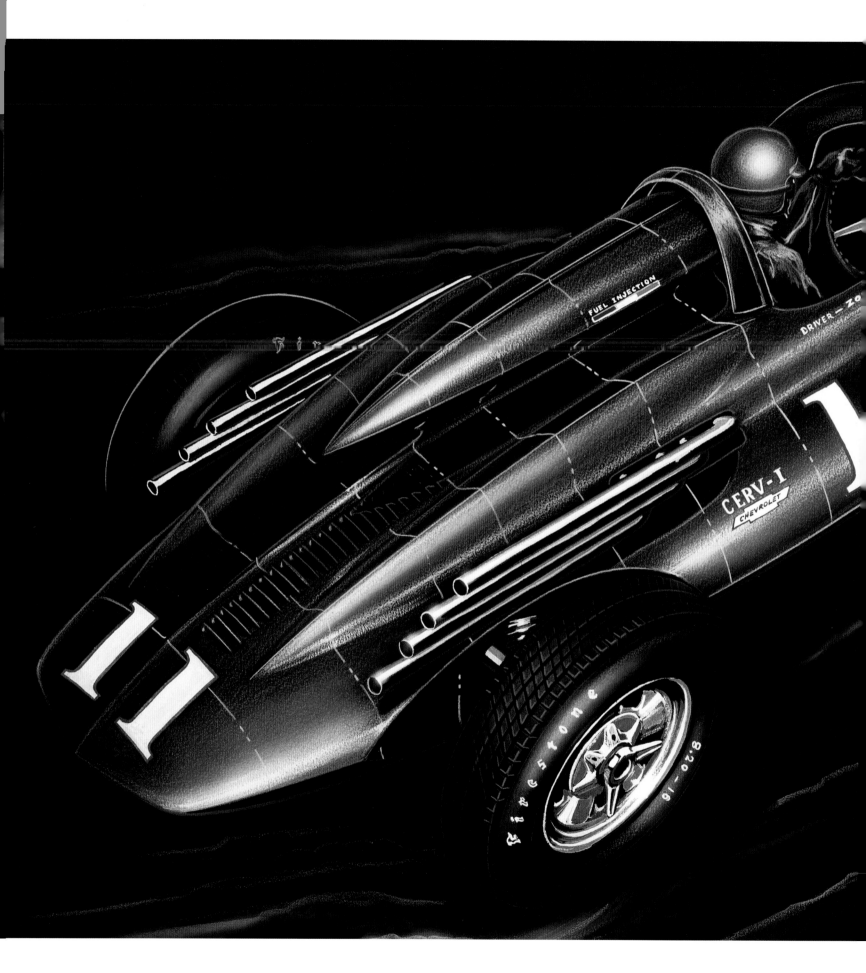

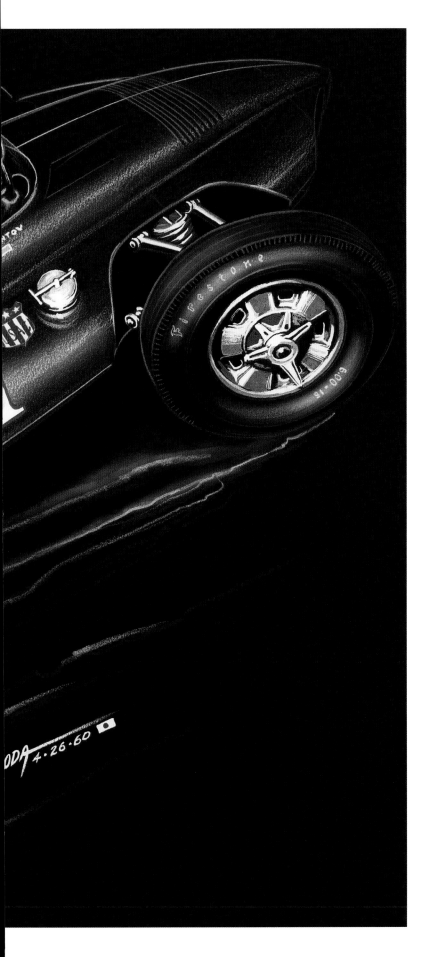

CERV I

While Arkus-Duntov had been sketching engineering designs of a mid-engine Corvette since 1957, his first working mid-engine machine was an open-wheel race car called CERV I. In selling the idea to his boss, Chevrolet Chief Engineer Ed Cole, he used a rationale concentrating on what the corporation might learn in developing daring new powertrain configurations.

His new car would be a test bed for all the best race car theory of the late 1950s in terms of powertrain, aerodynamics and using lightweight materials.

Thinking back to the monstrous, oversteering prewar Auto Unions, Arkus-Duntov wanted to dispel the myth that a high-powered yet tail-heavy car was inherently dangerous. "If you have an automobile powered like lawn mower," said Arkus-Duntov in an early 1990s interview, "low performance—it doesn't matter if you build the car tail-heavy or front-heavy. When you build high-performance automobile, it begins to matter. I establish as a guide 40/60 max."

With its sleek, narrow body, it looked like an Indy or Grand Prix car and helped pioneer using lightweight materials, as well as an independent rear suspension geometry eventually applied to the 1963 Sting Ray.

A small block V8 powered CERV I, the most exotic small block yet devised. Its lightweight aluminum core was made of a high-silicone alloy requiring no cylinder liners. The block weighed 90 fewer pounds than its cast-iron counterpart. Other components were made of lightweight magnesium. The engine was otherwise similar to the 1960 Corvette's stock 315-hp V8, with the same Arkus-Duntov cam, solid lifters and stock crankshaft, bearings, rods, pinions and rings. But breathing refinements allowed it to put out 353 hp at 6,200 rpm.

Larry Shinoda and Tony Lapine, up-and-coming designers who went on to do big things, designed the car's body under studio head Ed Wayne. With only two layers of fiberglass, the body weighed just 80 pounds. Its design allowed the car to squat at

Legendary designer Larry Shinoda's rendering of the 1960 Chevrolet CERV I concept car.

ABOVE, TOP TO BOTTOM: 1960 Chevrolet CERV I chassis; Arkus-Duntov with CERV I. OPPOSITE: Arkus-Duntov tests the CERV I at the GM Proving Grounds in 1959.

speed for even better aerodynamics. According to Arkus-Duntov, even the angle of the radiators contributed to higher downforce at speed, a fairly novel concept at the time. The CERV I's complete dry weight (without gasoline or other fluids) was a meager 1,450 pounds.

The machine also featured one of the first uses of a fuel cell in a race car. Designed by U.S. Rubber at Arkus-Duntov's urging, the cell was conceived to reduce the possibility of a fuel-fed fire. The rubber bladder was created to aircraft specifications, with baffles and a foam core. It was built to withstand rigorous tests, including a parachute drop from a high altitude with a full tank.

Arkus-Duntov again showed he was ahead of his time. After the fatalities in the 1964 Indy 500, in which Eddie Sachs and Dave MacDonald were killed in a fiery crash on the second lap, the United States Auto Club accepted Arkus-Duntov's recommendation that Indy cars require a fuel cell. A similar cell eventually became standard equipment in the 1975 Corvette.

The vehicle's transmission, a four-speed manual from the Corvette, was located in front of the differential. The differential featured a Halibrand casing with Chevrolet gears. While Arkus-Duntov would have preferred disc brakes, the technology wasn't there yet, so he went with a set of all-aluminum drums mounted inboard and flanking the differential. Like the Corvette SS, the car featured a variable braking control mechanism activated by a mercury switch.

As a forerunner of the 1963 Corvette Sting Ray, the rear suspension was independent and used the axle shaft as an upper link. "In passenger car and racing car, aim is identical," said Arkus-Duntov. "If you can find one piece to do what two pieces are supposed to do, that is good solution."

Nicknamed the "Hillclimber," CERV I was designed for both hillclimb competition, as well as flat-out speed. Arkus-Duntov had hoped to set a speed record up Pikes Peak in Colorado, but snowy conditions hampered his effort.

The Hillclimber had other potential uses. Arkus-Duntov would have loved to take it to Indianapolis, but its engine would have needed to be de-stroked to conform to the 205-cubic-inch limit for stock blocks then in force at the Brickyard. Had it been eligible, however, the car would have caused a stir because Indy was still

ABOVE, TOP TO BOTTOM: A Larry Shinoda design in profile; Arkus-Duntov takes the CERV I for a spin at Daytona Motor Speedway.

NEXT SPREAD The CERV II concept begins to take shape as this clay model in Chevrolet's Advanced No. 3 Studio.

the province of the front-engine roadsters, and Jack Brabham's revolutionary rear-engine Cooper-Climax didn't appear until 1961. Four years later, Jim Clark won the 500 in a rear-engine Lotus-Ford. Indy would never be the same, and Arkus-Duntov saw it all coming in the late 1950s.

In November 1960, Arkus-Duntov rolled out his creation on an international stage during the United States Grand Prix at Riverside, California. This event put Arkus-Duntov firmly on the map in the international racing community. Here, with the full backing of GM management, including engineer Walt Mackenzie and Cole, Arkus-Duntov demonstrated GM had a thing or two to offer the performance world. His audience included Dan Gurney, Brabham and Stirling Moss—the cream of the crop of Grand Prix drivers.

The presence of GM executives at the Grand Prix naturally led to intense speculation in the press regarding GM's racing intentions. For this reason, Mackenzie coined the name "Chevrolet Engineering Research Vehicle I," to avoid any overt connection to racing. The acronym didn't stop the inevitable press buzz, and Arkus-Duntov was recognized front and center as the brains behind the effort.

The press write-ups on the car were crafted to avoid the inference that GM was involved in racing. Even the track announcer's commentary was carefully scripted. Despite being an open-wheel race car, special pains were taken to keep CERV I separated from the "future Corvette" category in order to avoid hurting Corvette sales. Still, the masses were convinced this was a Grand Prix machine, the Automobile Manufacturers Association's ban on racing was about to be lifted and GM was going racing again. When the California race fans saw the blue and white racer from GM, they were sure.

Arkus-Duntov didn't alter that perception when he took some demonstration laps on Sunday before the race. He ran two laps at a steady 2:08. At this time, the Formula 1 lap record was 1:54.9, set by Moss—who proceeded to take CERV I around, shaving about five seconds off Arkus-Duntov's time. Brabham, Carroll Shelby, Chuck Daigh and Jim Jeffords also sat in the car just to get a feel for it.

Arkus-Duntov added some fuel to the fire, commenting to the press at Riverside about how motorsports involvement provides marketing credibility for a manufacturer. "Europeans can't be sold with slick advertising or cute slogans," said Arkus-Duntov. "They must see results. One of the best advertising devices in Europe is racing. If the Ferrari continually wins, the European reasons that it must be the better car."

The Riverside appearance garnered considerable publicity in car-enthusiast publications. It also resulted in a long story in Esquire magazine featuring a large color photo of Cole behind the CERV I's wheel. The article documented the transformation of the Corvette from its anemic roots at the original Motorama to a respectable sports car on the world stage, thanks largely to Cole and Arkus-Duntov. Although Arkus-Duntov would have loved to be the man in the picture, he was happy to share some limelight with Cole, if for no other reason than as a means of continuing Cole's support.

In an internal report to his bosses at GM, Arkus-Duntov concluded: "As an engineering tool, it gave us early reading on the functional performance of the 1963 Corvette suspension. It already gave us an indication as to how to bias our engine performance as we are heading toward higher power-weight ratios. Possibly, it indicates (the) type of tire our high performance 1963 Corvette would need. Since we are going to have two four-speed transmissions in the Corvette line, we have indication that a close-ratio, optional transmission should possibly get closer. As we continue to run the car, we probably will learn more."

After Riverside, Arkus-Duntov had the opportunity to expand CERV I's performance credentials and began to experiment with subtle bodywork changes and more horsepower. He wanted to conduct some high-speed runs, expecting to average more than 200 mph.

To this end, he fitted a supercharger and, later, according to author Karl Ludvigsen, twin turbochargers, bumping horsepower to a whopping 500 hp. Arkus-Duntov said the engine was so strong it lifted the front wheels completely off the ground under acceleration. With these changes, he made one last speed run at the GM Proving Ground in Milford, Michigan, where he lapped the 5-mile-long banked loop at a 206-mph average. He did so, not with the forced induction motor, but with a normally aspirated, bored and stroked 377-cubic-inch V8.

CERV II

Arkus-Duntov never forgot about going back to Le Mans again. Briggs Cunningham's production Corvette Le Mans effort in 1960, where John Fitch and Bob Grossman managed to finish eighth overall, served to revive Arkus-Duntov's dream of returning to France with a serious race car, a mid-engine car.

Going to Le Mans with a truly competitive, purpose-built car was easier said than done. GM was still in the midst of its racing ban, but Arkus-Duntov felt if he kept pressing the issue, he could appeal to the vanity of men like Cole, promoted to GM's head of car and truck operations, and new Chevrolet Chief Engineer Harry Barr. Ford and Chrysler had softened their anti-racing policies, and Arkus-Duntov reckoned it was only a matter of time before GM would do the same.

He began to think of a CERV I successor to serve his purposes. A Le Mans machine would require full bodywork, so Arkus-Duntov began working with designer Shinoda on a sports racing concept. In a Jan. 3, 1962, memo to Barr, Arkus-Duntov suggested using such a car in world championship endurance events, such as Le Mans and Sebring, both open to prototype and experimental cars of up to 240-cid displacement. Arkus-Duntov felt Chevy's money would be well spent going the prototype route because they'd relegate production-based GT cars to secondary status.

"Since no production numbers are required," wrote Arkus-Duntov, "we will see traditional European and American vehicles based on passenger car engines."

Arkus-Duntov proposed a prototype weighing about 1,500 pounds, powered by a 400-hp aluminum-block Chevy V8 with overhead camshafts. He also recommended aggressively using light metals—titanium, magnesium and aluminum—whenever compatible with the overall objective. The engine would be located ahead of the rear axle, and the chassis would be a space-frame or semi-spaceframe. Arkus-Duntov suggested a turbine or Wankel engine and was leaning toward the latter. "The Wankel engine is relatively new and is potentially a future powerplant for passenger automobiles," wrote Arkus-Duntov. GM later embraced Wankel technology in a big way, but at the time, the Curtiss-Wright Corp. held the licensing rights, and there were many unknowns about the engine.

Plans were drawn up for the prototype, including a manually controlled, power-shifted transmission and an engine featuring three-valve cylinder heads. The car was never built, though, due to renewed corporate pressure not to build cars that might go against the spirit of the racing ban. Arkus-Duntov shelved the project, but it was far from forgotten.

At the time, he had plenty of other things on his plate. Not only was he bringing his first complete Corvette model, the 1963 Sting Ray, to market, but also its racing variants, including the Z06 and the lightweight Grand Sport. Arkus-Duntov's plan was to build Grand Sport race cars, homologating them to become eligible for competition and thus navigating around the racing ban. Chevrolet could then sell them to customers.

Encouraged by then-Chevrolet boss Bunkie Knudsen's overall support for racing as a means to sell more products, Arkus-Duntov dusted off his CERV II plans. He wanted to be ready in case GM's policy ever changed.

This time, he had a few new tricks up his sleeve. He was now thinking about a prototype/race car featuring four-wheel drive and an automatic transmission in a full-bodied, mid-engine layout. While Knudsen never had official clearance to build a vehicle like this, he wanted a plan to be able to go up against Ford, which had decided to go full-bore into a "Total Performance" racing program including Indianapolis and Le Mans. Ford's Le Mans hopes rested on the Ford GT40, based on a British Lola sports-car chassis. CERV II once again skated on the edges of what Chevrolet could do under the AMA racing ban, but at least on paper it provided a formidable challenger in the Chevy stable.

Arkus-Duntov's interest in four-wheel drive wasn't a surprise. He had written a paper in 1937 called "Analysis of Four-Wheel Drive for Racing Cars" and had been closely observing what Auto

Arkus-Duntov takes demonstration laps in the CERV II racer, a car he'd hoped to race at the Le Mans 24-hour race.

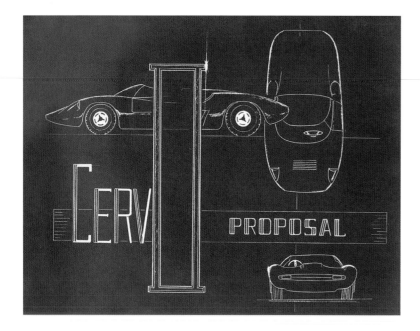

Union was doing in this area before the war. He was intrigued by the benefits of all four tires biting the pavement even if it meant carrying extra weight for the components necessary to drive the four wheels instead of two. Furthermore, a projected 550 hp from the engine was far more than could be used with the period's racing tires, according to Paul Van Valkenburgh, an engineer who worked for Chevrolet research and development boss Frank Winchell in the 1960s. Thus, Arkus-Duntov concluded four-wheel drive was mandatory.

Though four-wheel drive added traction and stability, Arkus-Duntov had to devise a means of distributing torque between the front and rear wheels. He elected to go with separate transmissions and torque converters for each end of the engine, reasoning that two transmissions would be lighter than one large transmission plus clutch, transfer case and driveshaft. It was a new principle, and Arkus-Duntov earned a patent on it.

Arkus-Duntov had more in mind than just basic four-wheel drive, however. He calibrated the torque converters to take advantage of weight transfer to pump more torque to the car's rear wheels under hard acceleration and less torque once it was moving at high speed. Arkus-Duntov also wanted the flexibility of multiple drive ratios, altering the bottom and top-end driving characteristics. He achieved this by equipping both axles with compact, two-speed gearboxes. Controlled by a single cockpit lever, they gave a direct drive and a 1.5:1 reduction.

Power came from an aluminum 377-cubic-inch V8 similar to the Grand Sport engines. The only difference was using a Hilborn constant-flow fuel-injection system instead of the Weber carburetors the Grand Sports used. During testing at the GM Proving Ground, Arkus-Duntov achieved a 0-60 run in 2.8 seconds, as well as hitting 214.01 mph on the test track.

CERV II also used some of the most advanced race tires in existence at the time, developed by Firestone. They were wide and low profile in a 9.5-by-15 size for all four corners, mounted to 8.5-inch Kelsey-Hayes magnesium wheels.

ABOVE, TOP TO BOTTOM: A rendering created for the AXP-789 program, a link between CERV I and CERV II; The CERV II had four-wheel drive and could hit 60 mph in 2.8 seconds—incredibly quick for its time; **OPPOSITE:** CERV II.

GRAND SPORT 2—XP-817

Quite simply, CERV II represented everything Arkus-Duntov wanted in a Le Mans-bound race car, but now he had to sell the corporation on racing it. That process involved competition from Winchell's Chevrolet R&D group in a sports racer called XP-817 or, simply, Grand Sport 2. Winchell had been closely but clandestinely linked with driver/engineer Jim Hall's privateer Chaparral Racing team in Midland, Texas.

Winchell cultivated the Chaparral relationship and appointed one of his most promising young engineers, Jim Musser, to be the chief liaison. In this mutually beneficial relationship, Chevrolet gained the use of the Chaparral shops and test track in the arid scrubland outside of Midland, Texas. Chaparral, in turn, received lots of hardware to play with from Chevrolet—engines, transmissions, suspension pieces and so on, all in the interest of research and development.

In fact, Arkus-Duntov found himself in a three-way internal battle for resources against not just Winchell but also Vince Piggins, who managed an engine group supporting NASCAR racing. Although the three departments never competed in the real world because the projects never went head to head, they did contend for resources and money. This was especially true between Arkus-Duntov's Corvette group and R&D, which had a bigger budget and more facilities, thanks to its involvement with higher-volume cars, such as the Corvair.

Having different approaches to their jobs, Winchell and Arkus-Duntov, in particular, were antagonists. Winchell might have been the better fit in a large corporation like GM. While extremely competitive, he didn't mind keeping a low profile, as opposed to Arkus-Duntov, who enjoyed basking in the limelight even if it meant irritating senior GM managers like Cole. "Frank was opposed to exposure," said Van Valkenburgh, the former Chevrolet engineer who wrote the book "Chevrolet = Racing ... ?" "He wanted Chevy R&D to be invisible, whereas Zora was the opposite—he wanted to be a public figure. Frank felt that the more

The Grand Sport IIb was developed with Jim Hall's famous Chaparral Racing team in Midland, Texas.

visible he himself was, the less he could get away with. Winchell was totally hands-on and self-taught. He didn't even have an engineering degree. He went to work for economic reasons. Everybody worked for Frank; nobody worked with him. He was strong-willed, assertive," Van Valkenburgh concluded.

Regarding Winchell, Ludvigsen wrote: "A tough, blunt customer who lacked a formal degree, Winchell was sometimes characterized as 'General Bull Moose after one of Al Capp's characters in his comic strip, Li'l Abner.'"

The Chaparral 2 developed into the GS2 (Grand Sport 2), which bore comparison to Arkus-Duntov's CERV II. Like CERV II, the GS2 had an automatic transmission that Winchell designed but had two-wheel drive instead of four. The GS2 name was borrowed from Arkus-Duntov's modified Grand Sport Corvettes and, according to Van Valkenburgh, Winchell might have thought it necessary to one-up Arkus-Duntov.

The GS2 was built at Winchell's shop and then trucked down to Midland, where it could be tested in absolute secrecy. The car was a beauty—a sports racer with an aluminum-block 327-cubic-inch engine and a single-speed automatic transmission. It was a forerunner of many generations of Chaparral sports racers that took American road racing by storm in the mid- to late 1960s. Chaparral was a leader in its aerodynamic research. Hall developed spoilers and articulating wings to keep his cars glued to the ground in both high-speed and tight corners.

Naturally, CERV II, the GS2 and the men who created them drew comparisons to one another. Each man criticized the other's efforts. Winchell was a powertrain guy who did not look kindly on Arkus-Duntov's twin-torque converter arrangement; Winchell thought it was too complicated for a race car. Arkus-Duntov, on the other hand, was an engine/suspension guy and thought Winchell didn't understand the relationship between the track (the distance between each set of wheels) and the center of gravity. He felt Winchell's cars were too narrow. Arkus-Duntov admitted that Winchell could irritate him, especially when not receiving adequate credit for the engines he provided to Winchell's group. But when push came to shove, Arkus-Duntov ultimately acknowledged that Winchell had some good engineering approaches, and the two were ultimately in a position to learn from each other. The steering knuckles used on the Grand Sport Corvette, for example, were from Chaparral.

Considerable budgets went into Winchell's programs, and, at times, Arkus-Duntov was envious. For the company's key decision-makers, directing more money into Winchell's programs was their way of controlling Arkus-Duntov's zeal for high-profile activities. Arkus-Duntov found it impossible to lie low, and Winchell excelled at doing so. If GM was going to go racing, it much preferred to fly under the radar with Winchell and Chaparral. Furthermore, Midland's Rattlesnake Raceway, home track for the Chaparral team, afforded much more secrecy for testing than the GM Tech Center.

CERV II VS. GS2

Eventually the cars were tested against each other. In March 1964, CERV II was brought down to Midland and tested against the GS2, but teething problems with the two-speed box driving the front wheels and issues with the disc brakes caused CERV II to suffer in comparison to the simpler R&D car.

As a result, CERV II never saw the kind of limelight on international racetracks Arkus-Duntov had envisioned. In summer 1964, the edict came down from Chevrolet General Manager Knudsen that CERV II would not be used to compete against Ford at Le Mans. Instead, Chaparral would carry the Chevy torch. A Chaparral 2F competed at Le Mans in 1967 and showed enough speed to have its overhead wing banned by the organizers the following year. During the race, the car, driven by Phil Hill and Mike Spence, succumbed to automatic transmission problems. Not being chosen to represent GM at Le Mans was yet another bitter blow to Arkus-Duntov, who lost a chance to race his most advanced car to date. To add insult to injury, one of his biggest rivals had outmaneuvered him within the halls of GM.

CLOCKWISE FROM TOP LEFT: An XP-817 rendering in GM Design's Advanced No. 3 Studio; Arkus-Duntov with an engineer; Chevrolet's Grand Sport IIb on display in a management review in GM's Design auditorium.

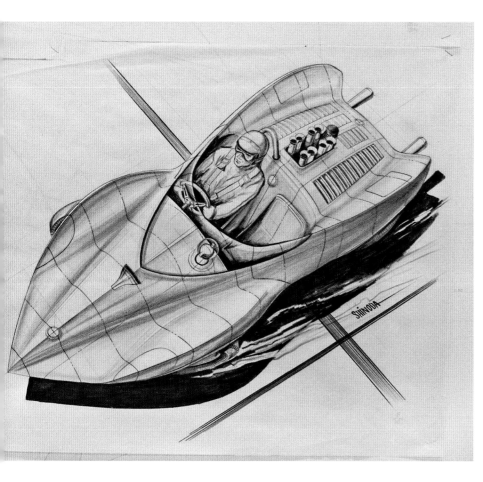

Several years later, the cars were matched again for tire tests, with CERV II being equipped with the aluminum version of the fabled L88 big block. Again, it went up against a modified GS2B. The R&D group was also anxious to test four-wheel drive against suction traction as the best way to make a race car go quickly around a corner: Chaparral at the time was working on the 2J, which used two motorized fans at the rear of the car and a rear wheel enclosed by a full skirt. The object was to create a low-pressure situation under the car, vacuuming the car to the pavement.

By the time CERV II was brought back to Midland, it was several years old. "I was doing the computer simulations in 1968 for suction traction with the Can-Am race series in mind," said Van Valkenburgh. At the time, Can-Am was the SCCA's top professional racing series. The cars were wide, low sports racers with no limits on engine displacement or aerodynamic devices. As such, the Can-Am was a racing engineer's dream series. "I was also looking at four-wheel drive as an alternative to suction traction and was therefore interested in looking at CERV II again. My thinking was that suction traction was far more effective than four-wheel drive."

During the tests, the GS2B with a 327 V8 outdid Arkus-Duntov's car by a good margin, according to Shinoda, who was there for the tests, in a 1997 interview. "It was several hundred pounds lighter," said Shinoda. That killed any hopes of resurrecting the CERV II program. "Zora took it pretty bitterly," said Shinoda. "He felt that the guys who were running the tests may have sabotaged his car."

Winchell first began to venture into prototype territory when he teamed up with Bill Mitchell on a Corvair-based project. Mitchell wanted a sports car based on the Corvair, and Winchell was working on various configurations with the Corvair driveline package mounted ahead of the rear axle. Musser was given the job of engineering Chevrolet's first monocoque chassis, and Shinoda did the body. The result was the very slick-looking Monza GT, or XP-777. The Monza GT caught Hall's attention. He invited Winchell and his staff down to Midland to look at a mid-engine V8 fiberglass chassis he was working on called the Chaparral 2.

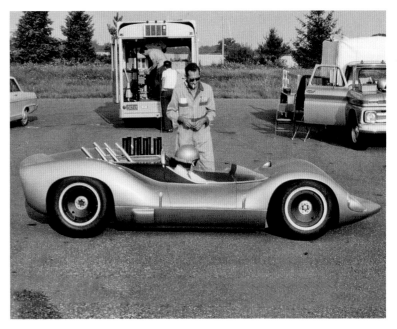

OPPOSITE: Jim Hall (center) gets ready to test drive the GS II at Chaparral's facility in Midland, Texas. ABOVE: The CERV II race car had two transmissions—one for the front wheels and one for the rears. The GS IIb was Chevrolet R&D boss Frank Winchell's idea.

ASTRO II—XP-880

As the third-generation Corvette was on the drawing board, Winchell tantalized the world with Astro II, also known as XP-880. Astro I, by the way, was a mid-engine concept based on the Corvair Monza GT and was also created by Chevrolet R&D.

Astro II was meant to be Chevrolet's answer to the threat of a road-going version of Ford's Le Mans-winning GT40. It followed on the heels of the rear-engine prototype XP-819. With its V8 engine located behind the rear axle, XP-819 proved to be far too tail-happy to be suitable for a production car. In contrast, Astro II, designed again by Shinoda, looked ready for production. Winchell assigned engineer Larry Nies the task of laying out the car, which became known as XP-880. The radiator was located in the rear. Air was also sucked out the back of the radiator via a large fan that ran when the engine was running. Power came from a stock 390-hp V8.

It first appeared at the New York International Auto Show in April 1968. But the car had an Achilles heel. Wrote Ludvigsen: "It used an automatic transaxle that was no longer in production, was too weak to take on the torque a high-performance version of the Mark IV could deliver and was poorly suited to a sports car with only two forward ratios. Nor were there any manual transaxles in volume production anywhere in the world."

Corvette sales continued to climb to their highest levels ever in the 1970s, and Chevrolet became more reluctant to mess with a good thing. At the time, there were still many drawbacks to mid-engine cars, including excessive cockpit noise and very limited storage space. High tooling costs also weighed heavily in the decision to stick with a front-engine layout.

But Arkus-Duntov kept his heart in the game and tried to engineer a mid-engine car that could be offered to customers at a non-exotic price. He also knew he would need to do it with existing parts to keep the tooling costs down.

GM Design's Jerry Palmer and Henry Haga designed a sleek silver body for the resulting machine, known as XP-882. It was low and wide, resembling the stance of a Ford GT40. The car appeared at the 1970 New York auto show.

The newest version of America's most popular sports car reflects, according to a 1970 Chevrolet press release, "the concern of designers at General Motors Styling and Chevrolet engineers for driver and passenger accommodation, effective air conditioning, luggage space and provision for an automatic transmission. Most mid-engine chassis layouts require more compromises than Chevrolet is willing to accept for a car with the Corvette's widespread appeal and enthusiastic following.

"The Corvette prototype is a design study that demonstrates the divisions' continuing research of mid-engine sports cars," wrote Chevrolet General Manager John L. DeLorean in the release. Despite all the hoopla, there were still no plans for production.

OPPOSITE: A rendering of the stunning Astro II, an idea meant to meet Ford's GT40 head-on. **ABOVE:** Rendering of the Corvette Astro II concept.

NEXT PAGE: Astro II interior.

XP-882

By the early 1970s, Arkus-Duntov knew he was running out of time—he had just five more years before his mandatory retirement age. Despite his proposals being shut down for a mid-engine Corvette in 1963 and 1968, he never lost hope. Now that he was chief engineer, having received the official title in 1968, he knew he had one last shot.

The main obstacle was the lack of proper mid-engine powertrain components. It was the same old story—tooling costs would run into the millions to supply the necessary automatic and manual transaxles. If Arkus-Duntov were to achieve his goals, he'd have to design a powertrain system using available parts within GM.

Arkus-Duntov's efforts centered on the experimental car known as XP-882. In conceiving this car, he had a wide range of GM drivetrain components to choose from. He found that hardware from the new front-wheel-drive Oldsmobile Toronado might be just what he needed. But there was a problem: Using these components would place the engine higher than Arkus-Duntov wanted it to be. It also limited him to an automatic-transmission version of the car because the Toronado didn't have a manual gearbox.

But Arkus-Duntov had a solution. He turned the engine sideways, mounting it transversely, locating the transmission on the side of the engine facing the passenger compartment and the differential on the other side. To connect the two, he ran the axle shafts through a tube in the center of the engine's oil pan. The differential case was then firmly affixed to the oil pan. Final drive gears were centrally located between the wheels in an aluminum housing. To add 4WD later would be a simple matter of adding a shaft to the bevel gears down through the center of the car to another differential. He earned a patent for this arrangement, granted in May 1971.

While transversely mounted engines are quite commonplace today, this was a relatively early application. The Lamborghini Miura and Ferrari's Dino 206/246 were among the few other cars on the market with this configuration. But the Miura had a unit

The XP-882 first appeared at the New York auto show in 1970—an era in which Corvette sales steadily climbed.

transaxle with a gear transfer drive at the left end of the engine and the transaxle entirely behind the engine.

Arkus-Duntov needed a simpler approach while striving for more compactness. His solution created a wheelbase only 95.5 inches long. By fully using the lateral space around the engine, he created more longitudinal space, allowing for a trunk. Clearly, he was working hard to chip away at the traditional objections to a mid-engine machine.

The finished XP-882 was low, wide and aggressive, with a roofline a full 10 inches lower than the pre-1963 Corvette. The car was designed in a coupe version only. Low-slung with that wide stance, it looked like a strong street-going competitor to the GT40, while maintaining a distinct Corvette resemblance with its sharp, chiseled edges and a new, louvered fastback roofline. The XP-882 weighed about 3,000 pounds, with a 44/56 front-to-rear weight distribution. The gas tank was located between the front wheels with a filler cap in the center of the front deck. The radiator was located up front, as well.

Concentric coil-shock units suspended the car at all four wheels, with a wishbone suspension and power rack-and-pinion steering at the front. In back, the independent rear suspension used the axle shafts as stressed members.

Although four-wheel drive was part of Arkus-Duntov's plan all along, it was not used in the original prototype. In an article published in Vette magazine, Arkus-Duntov wrote: "At the onset when the midship Corvette was authorized in 1968, I told the design group that we would plan the car with 4WD in mind at all times, but to keep mum about it. I felt that it was too much to ask upper management to have the foresight that 4WD for high performance sports cars was a necessary, coming thing. But the capability for 4WD was built into the chassis from the beginning. It was many months before I succeeded in convincing management that this was the way to go. The universal objection was over the complicated driveline."

The XP-882 took the world by surprise at the 1970 New York

auto show. General Motors had not planned to show the car, but since Ford was showing the mid-engine Pantera (built by de Tomaso in Italy), and American Motors was showing the AMX/3, Mitchell and Chevy Engineering Executive Alex Mair had second thoughts. Urged on by Arkus-Duntov, they elected to unveil the XP-882 in New York. In the rush to display the car, they didn't even have time to create a name—it was simply called "Corvette prototype."

But the Corvette prototype caused a sensation, completely upstaging the Pantera and the AMX/3. All of a sudden, Arkus-Duntov's mid-engine machine had momentum. Arkus-Duntov proceeded with developing a manual transmission in addition to the automatic. Plans were put in place to offer the car with a 400-cubic-inch version of the small-block V8.

Then the usual concerns began to creep in. While the XP-882 was a beautiful machine that would have been a fitting successor to the 1968 Corvette, there were worries at Chevrolet the car would be expensive to build, would offer no significant increase in performance and would be sold with only an automatic transmission. At the time, DeLorean, the Chevy general manager, was still considering the idea of a metal-bodied Corvette on the Camaro platform. A mid-engine machine would have been a much more expensive proposition.

OPPOSITE: Chevrolet said the louvered rear deck gave simultaneous engine cooling, as well as good rear vision and aerodynamics.
ABOVE: Arkus-Duntov shows the car to then-Chevrolet General Manager John DeLorean.

THE REYNOLDS ALUMINUM CORVETTE—XP-895

Arkus-Duntov kept the heat on. In 1974, he was so anxious to sell the world on a mid-engine Corvette that he was willing to forgo one of Corvette's most exclusive features since 1953: the fiberglass body. In its place would be a shining aluminum body designed by Reynolds Aluminum. Reynolds was anxious to showcase the possibility of aluminum automobile bodies and saw the mid-engine Corvette as a natural fit. A mathematical model was built to determine the necessary thickness of the aluminum skin—0.125 inch.

Arkus-Duntov received assurances that Reynolds would provide an aluminum body and structure identical to a steel car. The car, crafted in a small prototype shop in Detroit called Creative Engineering, wound up weighing 400 pounds less than an equivalent steel body. This finished car, called the Reynolds Aluminum Corvette, was a beauty. Its GM name was XP-895.

Given this convincing data and the extremely enthusiastic public reception for the XP-882 and the XP-895, it appeared for a time that a mid-engine Corvette was inevitable—whether its body was made out of aluminum or fiberglass.

XP-895, also known as the Reynolds Aluminum Corvette. The concept car's body broke with the tradition of Corvettes having fiberglass bodies, which they had since Day One.

ABOVE, TOP TO BOTTOM: Clay modelers work on the XP-882; stylists study the Corvette 2-Rotor. OPPOSITE: The 2-Rotor Corvette's engine was developed for small cars like the Chevy Vega.

THE 2-ROTOR CORVETTE—XP-897 GT

When Cole orchestrated GM's purchasing of the rights to the Wankel rotary engine, he began to seek a way to get the public excited about this radical new powerplant. A mid-engine Corvette would provide the perfect display case for a rotary engine.

So Arkus-Duntov and Cole agreed to create yet another new show car based on the XP-882. But Cole wasn't thinking only in terms of low-volume, mid-engine sports cars. He was really trying to sell America on the Wankel engine as a viable powerplant for high-volume small cars like the Vega.

Invented in Germany during the mid-1950s by Felix Wankel, the rotary engine was a whole different concept in internal combustion engines, and it showed immense promise. It was small and powerful and had fewer moving parts than a piston engine—and thus had potentially lower manufacturing costs. Cole was

sky-high on the possibilities. He had bought the rights from NSU/Curtiss-Wright to use the engine.

While the Corvette provided one obvious application, GM was eager to showcase the Wankel and built two different mid-engine concept cars to do so: The first was a smaller mid-engine coupe about the size of a Ferrari 246 Dino, according to Ludvigsen.

Designer John "Kip" Wasenko drew the car around the dimensions of a Porsche 914 chassis, with the Italian coachbuilder Pininfarina creating the body. The car was shown at the 1973 Frankfurt motor show.

While the resulting car was handsome and well proportioned, it lacked both the Corvette's brand character and support from both Arkus-Duntov and Senior Designer Chuck Jordan. There was also talk of making it into a successor for the Opel GT, but problems later surfaced and killed the car.

THE 4-ROTOR AEROVETTE

The second concept would be a much more powerful four-rotor Corvette. Cole approached Mitchell with the idea of freshening the XP-882's look to better showcase the new engine. The machine came to be known as the Aerovette. Under Mitchell's guidance, designers Jerry Palmer and Hank Haga created a sleeker version of XP-882, with an even cleaner almond shape (0.325 drag coefficient) and a smooth, flat fastback roofline with a window. Gullwing doors, inspired by the Mercedes 300SL, were another prominent feature. Arkus-Duntov loved the look: "In my mind, it was the best body that Bill Mitchell ever designed."

The Aerovette's four-rotor engine consisted of a pair of two rotor units connected together and mounted transversely, just like the original V8. According to Chevy engineer Gib Hufstader, "It was more than a case of hooking two two-rotor engines together. We had to handle cooling jackets, oil channels and oil pump. We tested it on dynos, and it was putting out between 360 and 370 hp. It became the largest automotive rotary engine ever built." The car's potential performance was incredible. Arkus-Duntov was quoted as saying, "This Wankel car is faster 0-100 mph than (a) 454," referring to GM's big block.

The plan was to show the Aerovette, as well as a two-rotor concept car with a Pininfarina body, at the 1973 Paris and London shows. All the GM executives from around the world were going to be there; Cole saw it as a chance to get everyone on board. The potential international press attention was also enticing.

But by the time the show cars were built and ready to display, problems with the Wankel engine were tainting its attractiveness to GM. Poor fuel economy, leaking seals and a tendency to run hot were all factors decreasing its desirability as a high-volume engine. Despite initially promising lower costs, a Wankel engine cost about $250 more to build than a V8—a huge differential. With the engine's many problems, suddenly linking the rotary engine to the Corvette wasn't so attractive. The rotary's demise also meant the demise of any mid-engine production Corvette.

ABOVE, TOP TO BOTTOM AND OPPOSITE: The 4-Rotor Corvette concept, or Aerovette, was faster to 100 mph than a big-block Corvette.

CORVETTE INDY

Still, GM never stopped exploring the midship concept. The stunning Corvette Indy debuted in 1986. Had it ever been built for the street, its wide, windswept look might just have stopped traffic. But its most exciting aspects were under the skin.

Powered by a version of the Chevy Indy 2,650cc racing engine, then competing in the CART PPG Indy Car World Series, the Corvette Indy was also a showpiece for active suspension technology and four-wheel drive. The active suspension system was developed by Lotus Cars of England, then a GM subsidiary. It used hydraulic controls activated by microprocessors reading the road. But instead of reacting to road inputs, it would actively smooth out the road as required, protecting occupants from the jarring of bumps and potholes.

For several years in the late 1980s, active suspension showed tremendous potential, and GM actually built several active suspension ZR1 Corvettes.

CERV III

Following the Corvette Indy's path, CERV III debuted at the 1990 North American International Auto Show in Detroit. Created under the direction of Don Runkle, a vice president of engineering, its technology included almost everything GM knew how to do at the time. Its body was carbon fiber, Nomex and Kevlar reinforced with aluminum honeycomb. It was fitted with a 650-hp twin-turbo LT5 V8 capable of 0-60 in 3.9 seconds, with a top calculated speed of 225 mph. Its body produced an exceptionally low drag coefficient at 0.277 cd. Like CERV II, it also featured four-wheel drive, thanks to a unique pair of automatic transmissions that produced six forward speeds.

OPPOSITE AND RIGHT: The Corvette Indy concept had for power a version of the racing engine Chevrolet was using then in the Indy Car World Series.

CORVETTE GTP AND IMSA PROTOTYPES

Any mid-engine Corvette history wouldn't be complete without mentioning the mid-engine race cars. The Corvette GTP competed in IMSA in the mid-1980s and was used by Chevrolet as a way to shift customer attitudes toward V6 powerplants. At the time, V6 engines were just entering the mainstream, and it seemed for a while a V6 Corvette might be a reality if gas prices and government regulations continued to increase. An opportunity to showcase V6 technology arose after IMSA created a new class for Grand Touring prototypes patterned after the European Group C machines that had been winning overall at Le Mans.

Chevy entered the fray with a special Eric Broadley-designed Lola chassis and turned it over to Rick Hendrick and his Hendrick Motorsports team to run a semi-factory effort with sponsorship from GM Goodwrench.

Palmer and Randy Wittine designed the beautiful black and silver Corvette GTP body and used many Corvette styling cues, especially the front profile and the large round taillights.

The car proved to be lightning-quick in the hands of drivers Sarel Van der Merwe and Doc Bundy, capturing many pole positions, thanks to its 1,200-hp turbocharged Ryan Falconer-built V6. But the Lola chassis couldn't handle that much power, and the car was only reliable enough to win two races—at Road Atlanta and Palm Beach in 1986.

In more recent years, Corvette-badged mid-engine prototypes with Corvette road car styling cues have raced in IMSA competition.

But these were cars built on one of several spec race car chassis authorized by IMSA and had little to do with advancing the evolution of a road-going mid-engine Corvette.

OPPOSITE: The Chevrolet IMSA GTP Corvette had a V6 engine producing a whopping 1,200 hp. **ABOVE:** The GTP car without racing livery.

PERSPECTIVE

It might be tempting to use the evolution of mid-engine Corvettes as a way to trace the development of the eighth-generation production car, but it's more accurate to say these cars kept the mid-engine flame burning.

Most of the mid-engine concepts never came close to production. They would have been too costly and impractical. These concepts represented the spirit and energy of the Corvette group, which always served as the tip of the spear for the Chevrolet Corvette—perhaps GM as a whole. It was a team spirit, initiated by Arkus-Duntov and extended brilliantly by successors such as Dave McLellan, Dave Hill, Tom Wallace and Tadge Juechter, that kept the Corvette in the middle of the conversation. They were part of an engineering culture that gave us technologies that became mainstream, from fuel injection to independent rear suspensions, active handling, traction control, carbon-fiber body panels, ceramic matrix brake rotors and magnetic ride-control systems.

One reason it took some 60 years to make a production mid-engine Corvette was that such a car had to be first and foremost a Corvette. That meant a familiar look, performance level, practicality and attainability. That has always been a key part of the Corvette franchise, and it explains the Corvette's being among the world's most successful sports cars.

Still, what made the mid-engine 2020 Corvette Stingray possible today? Perhaps more than the mid-engine concepts, it was all the things Corvette engineers learned in developing the previous-generation cars. The learning curve involved being able to build ultra-stiff, lightweight structures using some of the most advanced tooling technology on the planet. It also involved using everything Corvette engineers have learned in a 20-year Corvette Racing program and its numerous Le Mans wins. It was the ability to build a provocative car that wouldn't turn off the Corvette faithful with a lack of space or amenities. It was the ability to deliver astonishing performance at a price point right in line with traditional expectations. In short, it was the ability to keep the Corvette a Corvette.

BACK ROW, FROM LEFT: 1990 Chevrolet CERV III concept car, 1972 Chevrolet Reynolds Aluminum Corvette concept car; **FRONT ROW, FROM LEFT:** 1973 Chevrolet Aerovette concept car, 1959 Corvette Stingray racer.

THE MID-ENGINE CHAMPION

Corvette engineer Zora Arkus-Duntov made
it his life's goal to give the world a production
mid-engine Corvette. He finally got his wish.

Corvette engineer Zora Arkus-Duntov, the man considered the principal champion of a mid-engine Corvette, had a simple philosophy regarding production cars. It boiled down to this: Make the best race car you can, then let the chips fall where they may.

Racing was always his real love, his center. He firmly believed a great race car, particularly a mid-engine race car, would make the best sports car, despite some daunting challenges in terms of cooling, passenger comfort and practicality.

Corvette engineer Zora Arkus-Duntov made it his life's goal to give the world a production mid-engine Corvette.

GETTING HIS START

Arkus-Duntov joined General Motors in 1953 after seeing Harley Earl's Chevrolet Corvette concept car on a turntable at GM's Motorama at the Waldorf Astoria. His life would never be the same. It would be spent in the quest to make the front-engine production Corvette into a world-respected sports car and race car, while at the same time lay the groundwork for a future mid-engine production Corvette. It was a long and bittersweet struggle, but with the eighth-generation Stingray, Arkus-Duntov's dream has been achieved. His destiny to become the Corvette's patron saint began in an unlikely place: Russia. He was born in Brussels in 1909 to Russian parents who met there as exchange students. Raised by a Bolshevik mother, Rachel, in St. Petersburg, he witnessed the February 1917 Russian Revolution when he was 7. His family moved to Germany in the late '20s after Rachel was transferred there by the Soviet government. She had been a minister of culture under Lenin and boldly invited her lover, Joseph Duntov, to live with the family while Zora was growing up. Respecting both men, Duntov later added his stepfather's name to his own last name, creating Arkus-Duntov.

Unlike horse-and-buggy Russia, Germany was a place where Arkus-Duntov could truly feed his fascination with automobiles and racing. He attended several of Germany's finest technical schools and developed a particular interest in engines and super-chargers. He even wrote a lengthy article on supercharging, published by the noted German car magazine, Auto Motor und Sport.

Arkus-Duntov wasn't shy about approaching the racing giants like Mercedes-Benz and Auto Union to seek a position in their racing operations. Both companies were competing for state funding from Hitler's government. Arkus-Duntov was particularly fascinated by the mid-engine Auto Unions rocketing around racetracks with their supercharged V12 and V16 powerplants developing well over 500 hp. But they also had monstrous oversteer, causing drivers like Hans Stuck, Tazio Nuvolari and Bernd Rosemeyer to wrestle frequently with the cars around grand prix circuits.

Despite his attempts to get in the door at one of these giants, Arkus-Duntov had little chance at this stage of his life. His Russian-Jewish heritage would not fit well into companies charged with symbolizing Germany's growing nationalism.

OPPOSITE, TOP TO BOTTOM: Arkus-Duntov driving an MG, circa 1933–1935.
The Auto Unions fascinated him. **ABOVE:** In 1953, Chevrolet introduced
the Corvette at Motorama at the Waldorf Astoria in New York. Arkus-
Duntov joined GM after seeing the car there, changing his life forever.

Eventually Arkus-Duntov and his new ballet dancer wife, Elfi Wolff, were forced to move west to Paris after several threatening incidents in Berlin with Nazi storm troopers. His life took on an eerie calm before the storm—Wolff joined the Folies Bergère dance troupe and Arkus-Duntov built his first race car, a modified open-wheel Talbot with a supercharger he designed. He called it "The Arkus." It was an abject failure, but it taught Arkus-Duntov the disciplines and details in constructing a race car.

After World War II broke out, Zora and his family had to find a way to get out of Europe without exit visas. He was eventually able to get passage on a refugee ship out of Lisbon bound for New York. They landed at Ellis Island on a gloomy morning in December 1940.

The U.S. was at war a year later, and Arkus-Duntov had already found work using his German engineering education to consult for various companies gearing up for the effort. He eventually founded his own company to manufacture war munitions with his younger brother Yura. They called it Ardun Engineering, abbreviating their hyphenated last name.

When the war ended in 1945, the brothers shifted to the high-performance car market, building an overhead valve-conversion kit to increase the Ford flathead V8's power. The advent of overhead valve V8s, led by Cadillac in 1949, caused the business to eventually fail. Arkus-Duntov spent several years at Allard in the U.K. as an engineer and occasional driver. His first opportunity to race in the 24 Hours of Le Mans came in 1952, when Sydney Allard paired Arkus-Duntov with Frank Curtis in an Allard Motor Company works J2X. Arkus-Duntov soon grew tired of working for Allard's underfunded operation and its handbuilt cars.

Knowing Arkus-Duntov wanted to get back to America and work for a Big Three automaker, LeMay brokered a connection with Chevrolet Chief Engineer Ed Cole.

One thing attracting Arkus-Duntov to the Big Three was the fact that America had not sustained any damage to its infrastructure during the war. Thus there was a growing market for automobiles,

OPPOSITE: Arkus-Duntov and a Mercedes-Benz W165. **ABOVE RIGHT:** Arkus-Duntov's wife, Elfi, was a dancer with the Folies Bergère dance troupe in Paris.

NEXT PAGE: Arkus-Duntov racing the Allard J2X at Le Mans in 1952. Mechanical issues caused him to drop out.

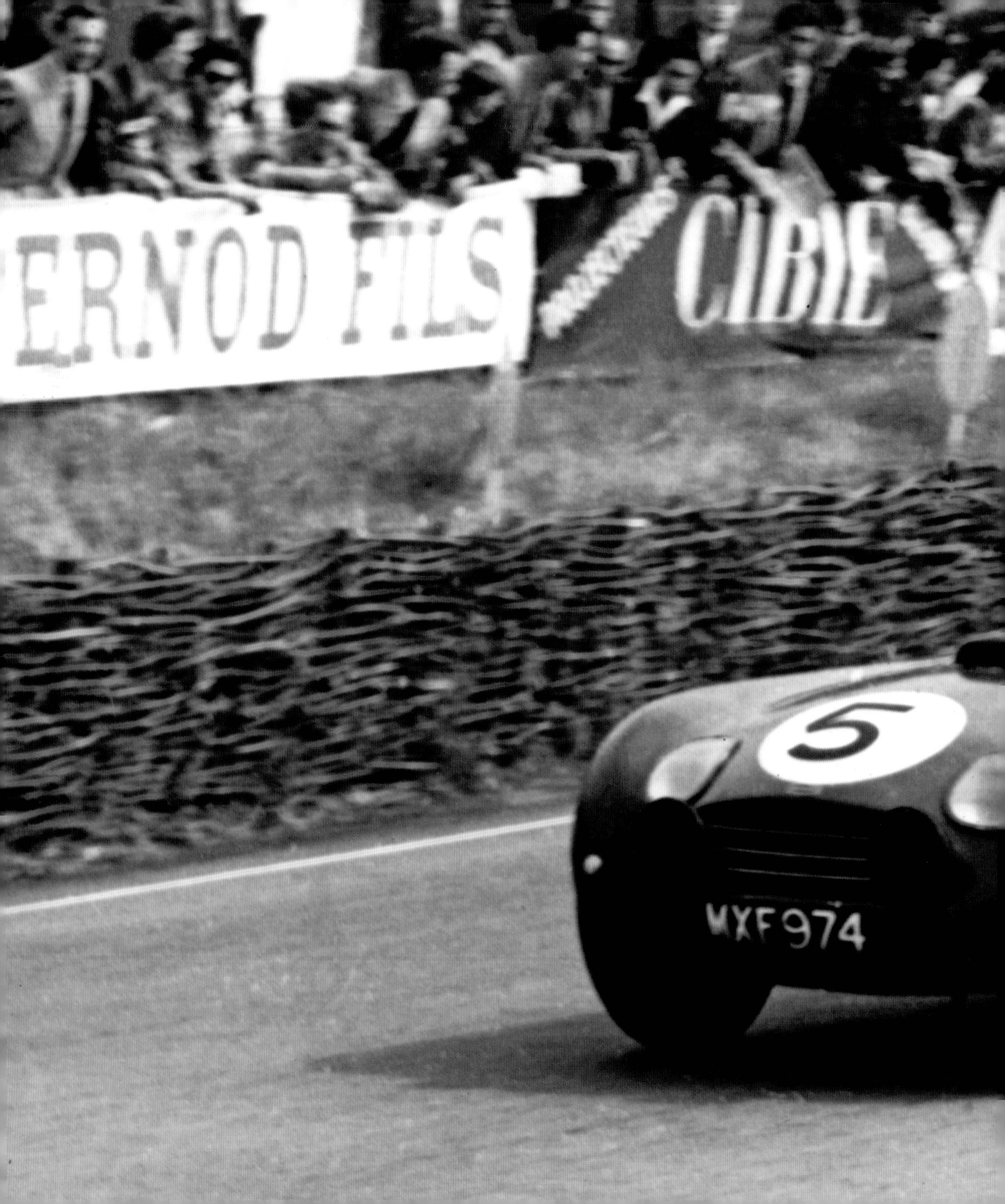
ERNOD FILS
CIBIE
5
MXF 974

along with tremendous resources to design and engineer them. He thought his racing background in Europe might allow him to make a huge difference for a country just cutting its teeth in sports-car racing.

Arkus-Duntov moved back to New York in 1952 and was working at Fairchild Aviation on Long Island when he attended the January 1953 General Motors Motorama in the Grand Ballroom of Manhattan's Waldorf Astoria hotel. This is when he saw Earl's Corvette concept car. Arkus-Duntov had no doubt in his mind that he wanted to work for GM.

He hurriedly dashed off a letter to Maurice Olley at Chevrolet Research and Development, leading to a job interview in Detroit. By May 1 of that year, he was officially a Chevrolet employee as an assistant staff engineer, reporting to Olley. In 1953, GM was still riding the crest of a postwar economy, and racing was not in its business plan—the company was too busy building millions of cars and trucks. Besides, American motorsports was in its relative infancy, not the big-time spectator sport it is today. The Indianapolis 500 was the only major event, NASCAR was barely six years old, and road racing was just getting started at places like Watkins Glen, Pebble Beach and Road America.

So it was with some consternation on Cole's part when Arkus-Duntov marched into his office and announced he had plans to drive at Le Mans, again for Allard. Cole was dumbstruck that his new hire had the audacity to ask for such a thing. Arkus-Duntov was so disappointed in Cole's response that he bought a one-way ticket to France, intending not to return to Detroit after the race. After dropping out of the race following mechanical difficulties, Arkus-Duntov returned to Detroit. His penance was to work on school bus drivetrains for the rest of the summer.

The tide was shifting at GM, though, and a year later, Arkus-Duntov was granted permission to race at Le Mans again, this time for Porsche. Arkus-Duntov sold the idea to his bosses based on the racing technology as well as air-cooled engine technology he might be able to bring back to GM. The latter was particularly important, with talk of a new Volkswagen Beetle competitor on the drawing board that would become the Corvair.

Porsche entered four 550 Spyders for the 1954 Le Mans race. Compared to its competitors, the 550s were feathery light, weighing only 1,200 pounds (dry weight without fuel or oil). Three entries had 1.5-liter engines, the fourth a 1.1-liter. All were air-cooled flat-fours with twin-overhead camshafts. The combustion chambers were hemispherical—just like Arkus-Duntov's Ardun engine—but contained two spark plugs per cylinder rather than one, an idea Arkus-Duntov later employed on a racing Corvette.

Arkus-Duntov drew the short straw and was assigned the 1.1-liter car. He co-drove it with Belgian endurance racer Olivier Gendebien.

During the race, it didn't take Arkus-Duntov long to notice how easy the Porsche was to drive. While there wasn't enough power for him to get too sloppy, he won his class and was impressed with the car's easy weight transfer, turn-in ability on dry pavement and traction in the rain. He filed those impressions away for future reference.

In fact, he produced a huge document analyzing the entire Porsche racing and engineering effort for the benefit of his GM bosses. The following year, 1955, Porsche invited him back, and he won his class again. He also witnessed the disaster wherein some 80 spectators were killed when racer Pierre Levegh's Mercedes-Benz careened into the stands after hitting Lance Macklin's much slower Austin-Healey on the pit straight.

GM, however, was just getting started. Cole was actually more aggressive than Arkus-Duntov, pushing for a full four-car factory effort at Sebring in 1956. Arkus-Duntov was concerned the Corvettes weren't quite ready. He didn't think the car's drum brakes were up to snuff for the 12-hour event. Cole, in turn, brushed right past Arkus-Duntov and brought in John Fitch from Mercedes to run the team.

Fitch conducted a credible effort in 1956, despite Arkus-Duntov's concerns, encouraging Cole to authorize constructing

Arkus-Duntov, driving a Porsche 500 RS Spyder, won the 1100cc class at Le Mans in 1954, finishing 14th overall.

ABOVE: John Fitch, left, with Arkus-Duntov and the Corvette SS, at Sebring, Florida. **OPPOSITE, TOP TO BOTTOM:** Ed Cole in the '56 Corvette SR-2 race car; the SS at Sebring.

a purpose-built race car—the Corvette SS—that could win overall, not just its class.

The actual race car arrived in Florida just a few days before the race. Resplendent in its metallic blue paint job, the car weighed 150 pounds more than the mule and had a new magnesium body, which turned out to be a heat trap for the driver. Making matters worse, the exhaust headers on each side wrapped around the cockpit, literally cooking the driver. Arkus-Duntov saw it as just another clear sign that future racers should have the engine behind the driver, not in front.

In the race, Fitch and Piero Taruffi got the driving duties, but without the development time the mule received, the shiny blue race car succumbed to a couple of mechanical gremlins after only 23 laps. It was a bitter end, but Arkus-Duntov was confident the SS could do well at Le Mans the following June. He received a major shock to his system when he heard GM was planning to withdraw from factory racing on the Automotive Manufacturer's Association recommendation. The AMA urged the Big Three to drop racing due to its inherent dangers after the 1955 Le Mans tragedy.

Nonetheless, the experiences helped Arkus-Duntov come to the conclusion that the engine being in the middle of the car, just ahead of the rear axle, was the best drivetrain configuration for both race cars and production sports cars. He thought weight shifted rearward improved traction, and with the engine behind the driver, forward visibility would be better. The flip side: A mid-engine configuration was noisier, cut cockpit room and compromised engine cooling because of awkward radiator locations.

Still, Arkus-Duntov thought sports car buyers would tolerate some inconvenience if the performance tradeoffs were strong enough and the driver—if no one else—could be well enough integrated into the machine. After all, a sports car was never supposed to be practical transportation, he reasoned. The rules are different. And in certain cases, sports car enthusiasts might regard the negatives as pluses—elements of a car's character and personality, adding to a car's not-for-everyone appeal. He saw a mid-engine Corvette as an opportunity to personally define the dreams of car enthusiasts on a worldwide basis—and to help establish General Motors as a performance leader and a volume

leader. Once again, he put aside reality: Most GM brass was more concerned with volume than with performance figures.

Despite the best efforts of Cole and the R&D staff, GM just didn't see itself as another Ferrari or Porsche. In addition to becoming a well-respected high-performance specialty car builder and a high-volume provider of reliable transportation, GM could create economies of scale, offering many more combinations of engines and drivetrains simply because the tooling costs could be amortized over many other car lines. And given its incredible resources, why shouldn't GM make the best performance cars in the world? To Arkus-Duntov, such a vision fit former GM Chairman Alfred P. Sloan's original notion of "a car for every purse and purpose."

The door was wide open in the late '50s, early '60s for GM to show leadership in this area because few manufacturers were working on mid-engine designs. Lamborghini's Miura didn't debut until the mid-1960s, and the de Tomaso/Ford Pantera and Ferrari 512 Boxer came during the late '60s and early '70s.

One thing working against Arkus-Duntov's mid-engine crusade was the front-engine car was starting to sell quite well and was in fact, following the Sting Ray's 1963 launch, making a profit for the

first time. Arkus-Duntov's superiors argued loud and hard that it would be foolish for Chevrolet to mess with a good thing.

However, there was one way the climate for Arkus-Duntov's thinking was reasonably favorable: While his bosses didn't want to necessarily do a mid-engine Corvette, they were aggressively pursuing new configurations for more mainstream passenger cars. Here Arkus-Duntov played a role in a radically different chassis/drivetrain concept Cole had in mind for conventional Chevrolets.

The concept, called the "Q Chevrolet," had the engine in the front, but the transmission was located in the rear. The transaxle featured an integrated starter motor. The brakes were mounted inboard to reduce unsprung weight. Arkus-Duntov was assigned to develop a Corvette version of the Q Chevrolet chassis. His Q Corvette design was compact, with a much lower center of gravity than any previous Corvette. To permit the lowest possible engine placement, a dry sump or separate lubrication system was specified, eliminating the oil pan.

Arkus-Duntov thought the layout provided subtle improvements but didn't think the transaxle shifted enough weight rearward to create any real performance advantages. "Like moving spare tire to rear of car," he said. The concept for the Corvette version of the car was later scrapped after Cole canceled the Q Chevrolet project.

Rather than accept defeat, Arkus-Duntov pushed to do a new mid-engine car. Some Q-car components still existed, and Arkus-Duntov reasoned he could make better use of them, placing the engine in back, just ahead of the rear wheels. Cole approved, so Arkus-Duntov and his team began to lay out his new vision. However, fewer suspension and drivetrain components were available from the Q Chevrolet than Arkus-Duntov had anticipated, and he and the GM designers could never agree on how such a car should look. Arkus-Duntov liked a short hood for better forward visibility, while designers, particularly design boss Bill Mitchell, favored a long-hooded look. These factors conspired to prevent building a running prototype.

ABOVE: A signed photo from John Fitch to Zora and Elfi, reflecting fondly on their time with the SS, circa 1957. **OPPOSITE, TOP TO BOTTOM:** The SS racer at Sebring; the XP-84 wood buck—the car led to the 1959 Stingray Racer and the 1963 production Corvette.

THE FABULOUS 1960s

It could be successfully argued that Arkus-Duntov's career reached a crescendo in the early 1960s. In quick succession, he developed two of the most important mid-engine concept cars in CERV (Chevrolet Engineering Research Vehicle) I and II, (see The Road to C8) plus he launched his first complete Corvette, the sensational 1963 Sting Ray.

GM Design was bursting at the seams with talent, with several young designers just itching to design a mid-engine Corvette. It's no surprise GM fueled a lot of the speculation over the decades. But in addition to the front-engine car now selling well, another factor standing in the way was the tremendous tooling costs required to change the Corvette's powertrain configuration.

That's why there was still a strong undercurrent of opinion within GM during the early 1970s to keep the Corvette as it was. Arkus-Duntov saw this firsthand at the London Motor Show when then-GM Chairman Richard C. Gerstenberg addressed the mid-engine issue directly with Arkus-Duntov. Wrote Arkus-Duntov, "Mr. Gerstenberg said like a father to a dim-witted child, 'We build Corvettes in numbers crowding 40,000 a year and still cannot satisfy the demand. Why do you insist that we abandon this successful design and barge head-on into a new design?'" Gerstenberg suggested a more cautious approach, indicating that if the market got soft, he would reconsider his position on mid-engine Corvettes.

Arkus-Duntov later wrote in Vette, "The words of Mr. Gerstenberg fell like a sledgehammer on my head. What about technical leadership for Chevrolet? What about performance leadership and my desire to produce a ne plus ultra American sports car? I felt I was betraying the people in the Corvette group, the Chevrolet people of Design Staff and Corvette people at large by withholding the introduction of the ultimate Corvette."

In part, he blamed himself—or, more specifically, his accepting a rotary engine in the latest mid-engine concept, known as the Four-Rotor and later the Aerovette, which was created in 1973. "I

LEFT, TOP TO BOTTOM: Arkus-Duntov in the CERV I racer at Pikes Peak in Colorado; the 4-Rotor engine in the Aerovette. **OPPOSITE:** Arkus-Duntov and the 1963 Corvette Coupe at the GM Tech Center in Warren, Michigan.

reproached myself for tying the Corvette with the rotary engine's thermodynamic inefficiencies. How to tell Mr. Gerstenberg that we liked reciprocating engines and that the rotary engine was an abortion enforced by Mr. Cole! No, I couldn't do that. I entered into an agreement with Cole, fair and square, so I had to bear the responsibility for the unholy alliance with the rotary engine."

While the Aerovette project was shelved, speculation Chevrolet would offer a mid-engine Corvette continued to run rampant. The question was not whether it would happen, but when.

A cover story in Road & Track in February 1977 predicted that for the 1980 model year, Chevrolet would offer a more conservative Aerovette, sans the Wankel engine and the gullwing doors.

The magazine speculated it was about time Chevrolet came out with the real thing after years teasing the public with several mid-engine show cars. The article cited the sales intrusion being made by cars like the Pontiac Trans Am, following the Corvette's enervation due to new fuel economy standards and emission controls.

Indeed, suddenly the Trans Am's performance was not that different from the Corvette's—at a far more attractive price. In 1977, the Corvette's base price was $8,647.65, and it was easy to bump the sticker to $11,000 after checking the option sheet. The Trans Am could be had for many thousands less. Given the narrowing gap, it was time for the Corvette to once again reassert its position as GM's technological flagship.

But the pundits were wrong. Corvette sales were clipping along at record numbers in the 1970s, and Chevrolet had no intention of tampering with a solid sales winner—the Corvette was a victim of its own success. The more profitable it became, the less daring it needed to be in terms of engineering and technology, and its role as a halo car diminished.

A collection of sketches of mid-engine Chevy Corvette concepts, all done by Allen Young and Randy Wittine.

NEXT PAGE FROM LEFT: A 1967 Chevrolet Corvette Sting Ray Coupe, 1967 Ford GT40 race car and a 1968 Chevrolet Astro II clay design proposal, all posing at the GM Design Patio behind the GM Design Center.

CLIMATE CHANGE

During his remaining years at Chevrolet before his retirement in 1975, Arkus-Duntov had explored several major technological enhancements for the third-generation Corvette including using a transverse fiberglass leaf spring (eventually adopted in 1982), antilock brakes (offered in 1985), a Bosch fuel-injection system and even an aluminum body.

Clearly, Arkus-Duntov helped set the agenda for automobile technology we take for granted today—yet in his mind, he never realized the ultimate Corvette, the car embodying everything he knew as an engineer.

As Arkus-Duntov's career wound down in 1973-1974, he made peace with Cole even though Cole was up to his eyeballs in crisis management. Cole's baby, the Corvair, finally died in 1969, a market failure after years of negative publicity, and a motor mount Cole had designed was subject to a massive recall. The Vega was having problems, and Cole faced a host of challenges regarding emissions and engines. These and other problems might have prevented him from reaching the top spot—chairman of GM's board. Unquestionably, the last few years' troubles had taken a toll on Cole, who in 1974 told Newsweek that after 44 years at GM, the car business wasn't fun anymore. He retired at the mandatory age of 65 on Sept. 29, 1974.

While many historians have characterized Cole as a staunch Corvette defender, his support, according to Arkus-Duntov, came not because he was a sports car enthusiast, but because he understood the halo effect a car like the Corvette could have on the rest of his Chevy division. "His interest was not Corvette at all, but passenger car," said Arkus-Duntov. "That's where the profits were, and he did not relate to Corvette, per se." Arkus-Duntov might have been overly critical on that score, though, considering how often Cole went to bat for Arkus-Duntov over the years as well as how much Cole enjoyed driving Corvettes.

THE DELOREAN CHALLENGE

Weeks after his retirement from GM in 1975 at the mandatory age of 65, Arkus-Duntov joined former Chevrolet boss John DeLorean as a consultant on DeLorean's new DMC-12 sports car project. The opportunity to join the DeLorean Motor Co. was irresistible. It held great promise because Arkus-Duntov knew DeLorean had to pursue unconventional approaches to distinguish his vehicle in the sports car market. Arkus-Duntov recognized teaming up with DeLorean might provide him with a way to fulfill his mid-engine dream, even if it wasn't for GM.

Such was DeLorean's belief in Arkus-Duntov that his original notion was to acquire the rights from GM to produce Arkus-Duntov's stillborn mid-engine prototype designs. The 1973 energy crisis, however, convinced DeLorean he had to pursue something lighter than Arkus-Duntov's concepts. He also knew he'd have to adopt a smaller displacement engine. DeLorean wanted Arkus-Duntov on his team and put him on a $1,000-per-month consulting retainer.

Much to Arkus-Duntov's disappointment, he never had a chance to design a car from the ground up for DeLorean. He spent several years on the sidelines as a sounding board and paid critic, but he never had any real influence on the designs.

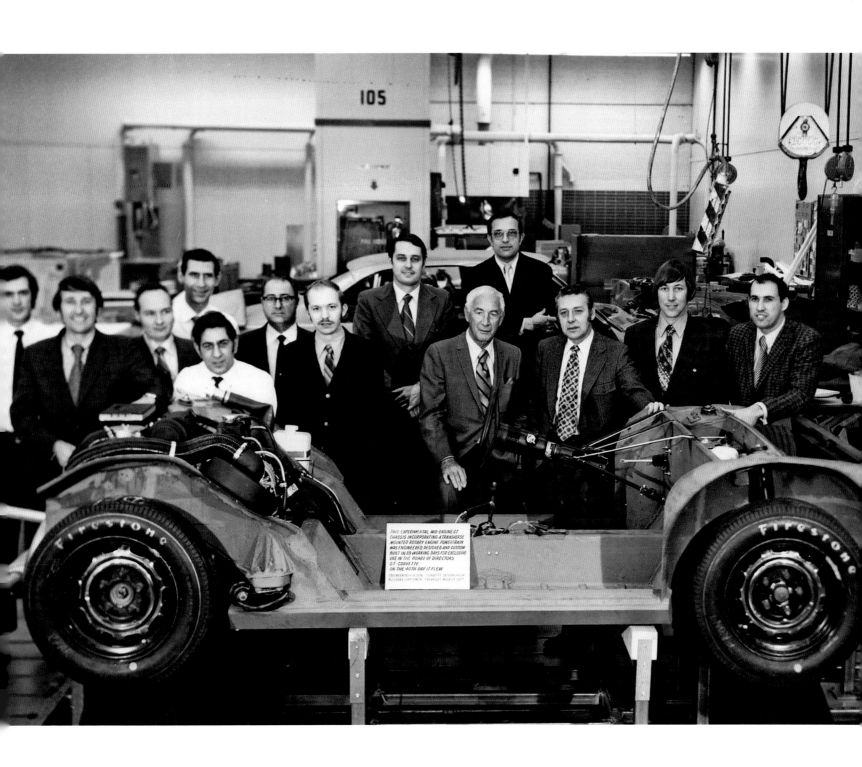

THIS EXPERIMENTAL MID-ENGINE GT
CHASSIS INCORPORATING A TRANSVERSE
MOUNTED ROTARY ENGINE POWERTRAIN
WAS ENGINEERED DESIGNED AND CUSTOM
BUILT IN 39 WORKING DAYS FOR EXCLUSIVE
USE IN THE BOARD OF DIRECTORS
GT CORVETTE
ON THE 40TH DAY IT FLEW.

ENGINEERING + DESIGN : CORVETTE DESIGN GROUP
BUILDERS-CRAFTSMEN : CHEVROLET MOCK-UP DEPT.

OPPOSITE: Chevrolet Chief Engineer Ed Cole speaks at his retirement in 1974. **ABOVE:** Arkus-Duntov, front row center, pictured with the Corvette design group and the 2-Rotor Corvette experimental car.

ZORA ARKUS-DUNTOV
ENGINEERING CONSULTANT
621 LOCHMOOR BOULEVARD
GROSSE POINTE WOODS, MICH. 48236

——

(313) 884-0748

this letter was sent approximately in June '91

Mr.J.C.Perkins
General Manager
Chevrolet Motor Division
Chevrolet Motors Corporation
30007 Van Dyke Ave
Warren, MI 48090

Dear Jim:

Everything repeats itself.

In 1955 only 700 Corvettes were sold (down 4000 from the previous year 1954) and cancelling of Corvette as a whole was immenent.

Today we are facing declining sales and thoughts are surfacing "what about cancelling Corvette as a whole.

Whithout belaboring the consequenses of this decision let's look at a way out of the predicament.

Sorry to say,but the current car was outdated by inseption in 1984. Front engine.rear drive,outdated configuration for the sports car.

I propose:

All engineering expenditures connected to the current car will be eliminated, instead what is needed,all wheel drive sports car with favorable mass distribu- tion in the order of 45/55 midship design and torque split accordingly with Viscous coupling.

We built two experimental cars,Corvette Indy and last "Corporation experimental car".

Both of these two cars incorporate driveline of the Midship Corvette of 1970-73.

Midship Corvette had only driveline to the rear wheels but was <u>designed to accept 4wd</u>

Universal uneasinee to the driveline was pocking through the oilpan.

Lo and behold that feature is contained in the '91 Acura Legend LS.

(G.M. patent attorney,R.J.Outland informed me that the live of a patent is 17 years. The patent was issued in 1969 and therefore is free for all).

We have two Midship cars - Aero Vette and all aluminum car. The all aluminum is a true representation of a production car.

The last thing I know R.D. has it.

The car was designed to have 350 cid and 454 cid racing engine with torque up to 600lb/ft and therefore all of the drive train was designed to take it.

Almost 15 years later the production car racing is non existant,no ink is generated by participation in this event.

Ink is reserved for cars having specialized chassis and engines.

Therefore effort to produce drive train for 350 cid engine only.

Steel chassis and fiberglass body car weight 3000 lb.

Using all light suspension component front and rear,exept the front knuckle will be carried over from the current car.

On a 95.5" wheelbase 45/55 weight distribution,4wd,spacious interior,exhaust piping catalystic converter in the back of the passager compartiment -soluti is timeless.

c.c Ralph Kramer

ABOVE: Arkus-Duntov's letter to then-Chevy General Manager Jim Perkins. OPPOSITE, FROM LEFT: Dave McLellan, Perkins, Corvette Chief Engineer Dave Hill and Arkus-Duntov at the National Corvette Museum in Bowling Green, Kentucky.

NEVER SAY DIE

Despite indifference on GM's part, Arkus-Duntov continued to volunteer his services as a consultant well into the 1990s. Not knowing where else to go, he became a regular visitor to then-Chevy General Manager Jim Perkins' office. As far as Perkins was concerned, these were courtesy visits. He tried to use Arkus-Duntov as a resource, if only out of respect and admiration, even though Arkus-Duntov's specific input was unlikely to be transferred to the Corvette platform group.

For his part, Arkus-Duntov never quite understood that Perkins had little control over the configuration of future Corvettes. That was now the Corvette engineering group's responsibility, a totally separate entity from Chevrolet ever since the late 1970s. "I sat with Zora for hours," said Perkins, "and had him work me over because he didn't understand that I didn't have the same latitude during my tenure as Chevy GM (1989 to 1996) as Ed Cole did. It was tough for him to understand the differences in GM from Cole's day."

But Arkus-Duntov never gave up hope. In a 1992 letter to Perkins, Arkus-Duntov proposed that all engineering expenditures connected to the current car be eliminated. Instead, he suggested a midship all-wheel-drive sports car with favorable mass distribution on the order of the 45/55 midship design and torque split accordingly, with viscous coupling. He pointed out that his patented driveline from the Aerovette, which ran through the oil pan, had since been adopted in the then-new 1991 Acura Legend LS after Arkus-Duntov's 17-year patent expired.

Several years later, Arkus-Duntov visited Perkins at Chevrolet just as final plans for the fifth-generation Corvette were to be locked down. It was clear before the meeting that the ship had already sailed as far as the new car's chassis and drivetrain were concerned. It would be front-engine, rear-drive, just like its predecessors.

"He came to my office, and a shouting match almost developed," said Perkins. "He, of course, thought mid-engine was the way to go and had hand-drawn plans for the C5 with a midship engine. He began to get angry, and I said, 'You don't understand.' Then he quickly stood up and stated, 'Then I build the son of a bitch myself.'

"He wouldn't speak to me for a while," added Perkins. "But a month or six weeks later, he came back with a proposal to run with some backers who were going to do a car. Who was I to say he was too old? That to me was the spirit and the heart and the soul of the man ..."

It was also a strong indication that 20 years after retiring from GM, Arkus-Duntov was still trying to control the direction of future Corvettes. He died in April 1996 without ever seeing a production mid-engine Corvette on the streets. Another 23 years passed before his dream was accomplished and the wraps were taken off the 2020 mid-engine Corvette Stingray.

GM President Mark Reuss, who had the honor of driving the car onto the stage at its July 2019 launch at California's Tustin Air Station, was clearly thrilled a production mid-engine Corvette happened on his watch. Reuss had been instrumental in getting the CERV cars back into GM hands after decades of private ownership and spoke that evening about visiting the Chevy engineering shop at the GM Tech Center with his GM Executive Engineer father Lloyd Reuss and meeting Arkus-Duntov himself in the shop.

Asked what he would say to Arkus-Duntov now, Reuss simply replied, "Hey, we got it done. You started it and we finished it. And we had a great time along the way."

Arkus-Duntov, here in a 1990 Corvette, lives on through the mid-engine C8, his dream coming to fruition more than 20 years after his death.

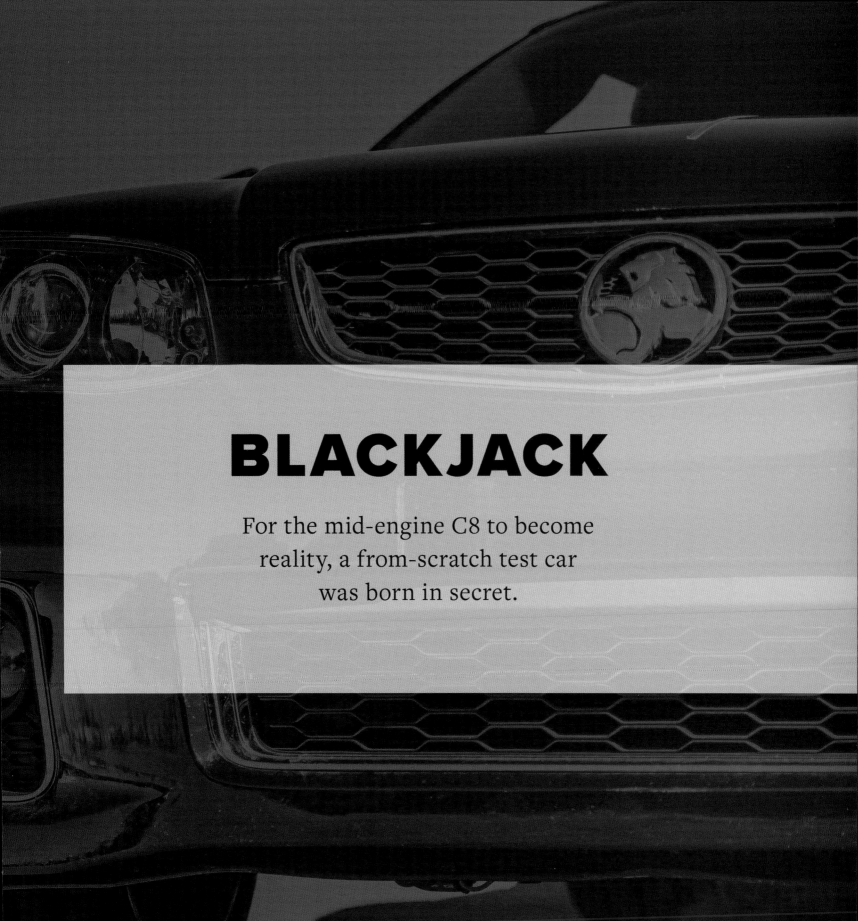

BLACKJACK

For the mid-engine C8 to become
reality, a from-scratch test car
was born in secret.

How do you begin to create a mid-engine Corvette when none existed before? That's the question GM engineers had to answer when they began development of the 2020 Corvette Stingray. Once the mid-engine Corvette program was approved by GM management and the development team was assembled, they needed to set the parameters for the car in terms of performance. They benchmarked competitive cars and began to formulate an idea of how the Corvette could be better. In order to validate the goals they set for the new car and their ideas about how to achieve them, the team needed to build a test vehicle. The very first crude iteration of the C8 Corvette, probably one of the best-known vehicle prototypes in history, would come to be known as Blackjack.

The Corvette development team had to find an innovative way to keep prying eyes and spy photographers from realizing the mid-engine configuration was on its way—they used a Holden Ute to disguise the project.

OPERATION: BLACKJACK

At General Motors, early development vehicles and concept vehicles are built by the Advanced Vehicle Integration (or AVI) team at the Warren, Michigan, Tech Center. Building Blackjack would fall to them and end up being the most challenging build that they'd ever done.

"Usually a vehicle like this has some type of donor property that you're using the structure from or exterior panels from," Corvette lead development engineer Mike Petrucci notes. But the mid-engine C8 required an all-new backbone body structure that could accommodate a mid-mounted engine. Because that structure was fundamentally different from that of the front-engine C7 Corvette, this test mule's mock-up of that architecture would have to be made from scratch. AVI engineers created a facsimile of the intended body structure, milling its elements from solid aluminum—a fantastically expensive and labor-intensive process that would never work for mass production. (In the real car, the underlying structure would be created from high-pressure die castings, a change from the hydroformed aluminum used in the previous Corvette.)

Starting from scratch was hard enough, but doing everything in secrecy compounded the project's difficulty.

Naturally, at this early date General Motors did not publicly acknowledge that it was working on a mid-engine Corvette. (That announcement wouldn't come until April 2019, just three months before the finished car's reveal.) The upcoming C8's mid-engine configuration wasn't just kept from outsiders, but from as many people as possible within GM, as well. AVI, then, couldn't simply build this mule in the middle of their workspace, among all the other vehicles they were working on. The AVI team had to create a separate, access-controlled room within their building, where they could work on the top-secret C8 development vehicle.

RIGHT: Code-named Blackjack, the development mule never found its way to public roads; rather it was only used at GM's Milford Proving Grounds to determine if the mid-engine configuration would work, with a majority of the test work done under the cloak of darkness.

NEXT SPREAD: Besides validating the basics of the mid-engine configuration, Blackjack also played a key role in the understanding of how air would flow through the body to keep the engine and transmission operating at peak performance.

THE PICKUP PUT-ON

The team also had to look ahead to when the prototype was complete and ready for testing. To roll Blackjack out of their shop, they needed bodywork that wouldn't be identifiable as a mid-engine Corvette prototype. The solution required some next-level creativity and inspiration from GM's Holden Division in Australia—the mid-engine sports car would masquerade as a pickup truck. Or more accurately a Holden SS-V ute—a passenger-car-based pickup truck in the style of the Chevy El Camino. The team had also considered other solutions: a station-wagon shape using blacked-out rear windows, as well as a small van, but they settled on the Holden ute.

The C8 Corvette chassis and running gear, however, did not fit easily under the front-engine, rear-wheel-drive ute's body. Far from it. The Aussie pickup is a foot-and-a-half longer and has a nearly 12-inch-longer wheelbase than the C8. That meant none of the Holden's factory body panels could be used. The only Holden exterior elements that could be carried over intact were the front fascia, the headlights, the taillights and the side-view mirrors. All of the body panels had to be custom-crafted by AVI—mostly from fiberglass, some from carbon fiber—to fit over the Corvette test mule's structure and mechanicals. That included the hood, the front fenders, the rear quarters, the faux tailgate and the tonneau cover that hid the engine. As Petrucci characterizes the result, the Blackjack "didn't use the body of a Holden ute, it used the theme of a Holden ute."

The test mule had large fender flares to accommodate the C8's wider front and rear tracks and its wider wheels and tires. A huge panel covered what would be the open rear bed in the truck, with the mid-mounted engine hidden underneath. A giant wing sprouted from the rear of that tonneau panel, and the two towers that supported it (which were made of wood) contained openings at the front that ducted cooling air to the engine below. The wing itself was shaped to provide aerodynamic lift—the opposite of what a rear wing is usually intended to do. That's because the mule's nose also had lift at speed, and the important thing for testing was that the front and rear ends be balanced aerodynamically

Although you might not guess by looking at it, the Blackjack actually contains a considerable amount of C7 Corvette. Whereas the SS-V had a more upright, sedan-like glass area (the model is based on the Holden Commodore sedan), the Blackjack's roof is lifted straight from the C7, and the windshield header was also a C7 piece. The prototype's interior was largely that of a C7, and the outer door skins were custom-made to fit over C7 door inners.

The mid-mounted engine powering the Blackjack was a Chevy LT1 V8 from a C7, whereas the real C8 would get an LT2 dry-sump V8. Also in contrast to the production C8, the test-mule engine was mated to a ZF seven-speed dual-clutch automatic rather than the Tremec eight-speed unit that would ultimately be tapped for the production car. The C7's electric modules also were donated for use here, as were its cooling and brake systems. The Blackjack took some eight months to build, which is nearly twice as long as usual.

The franken-ute's nickname itself was a further element in keeping the mid-engine project a secret. As Petrucci explains, "It was a codename that allowed the AVI team to work on the vehicle, and have emails going around talking about the vehicle, but created a sort of ambiguous cover for it in case your cube-mate looks over your shoulder and saw you reading an email."

The only off-the-shelf pieces used in Blackjack were the Holden Ute front and rear fascias. The structure was built from billet aluminum, and the cockpit was borrowed from a C7 Corvette. Since the DCT transmission used in the 2020 Corvette hadn't existed, the team used a PDK transmission.

THE BLACKJACK'S MARCHING ORDERS

How could the Blackjack contribute to the team's learnings about the new Corvette when so much about the vehicle was not representative of the new car? To answer, Petrucci gives an overview of the car's development and what is required from the test mules at various points in the process. "At the beginning of a vehicle development project, there are very coarse learnings that don't require things that are specific to the particular project," he says.

The focus when the team was working with Blackjack was around body structure targets. Blackjack also had a fair amount of suspension adjustability, with bolt-on components in the body structure that allowed engineers to move the attachment points for upper and lower control arms and evaluate a variety of setups. Those hard points are determined at a very early stage, and once set, they don't change later in development during fine-tuning.

"When you're trying to determine attributes like that," Petrucci says, "you don't care what the car looks like, what the aerodynamics are, how much horsepower it has. You care about everything from that tire contact patch all the way up to the steering wheel. You also care how much the vehicle weighs. Those were all attributes that we made sure that Blackjack accurately represented."

Thus, when Petrucci and his team sat down with the AVI folks to scope out the build, there was a whole list of elements they held sacred, aspects that needed to be adjustable and items they did not care about. Among the elements that had to correctly represent the C8 were the overall vehicle mass, the car's front-to-rear weight distribution, the wheelbase, the front and rear track, and the position of the occupants relative to the wheels. Items that needed to be adjustable included the suspension attachment points and even its configuration, the wheel offset and sizes, and the tire sizes. The remaining elements the vehicle development engineers did not care about, and in those areas AVI was free to take whatever liberties were necessary to make the prototype work and drive and to fit the sort-of-Holden-ute configuration.

THE BLACKJACK GOES TO WORK—IN SECRET

In all, Blackjack served for roughly two years. And it had different goals over that period. The early objectives were figuring out body structure targets and suspension kinematics, including the hard points and geometry. Later on, it was used to help get the engine sound characteristics to where the team wanted them, an area that was particularly important to the Corvette. The exhaust manifolds, the catalytic converters and the rest of the exhaust system were initially designed virtually; that engineering had to be confirmed with testing of the physical hardware. That was done on Blackjack, with the LT1 engine fitted with LT2 headers. The testing resulted in changes to the components to perfect the sound of the new Corvette.

Although the vehicle was wearing a Holden ute disguise, General Motors still did not want the Blackjack to be seen during testing. GM's Milford (Mich.) proving grounds, however, is a large facility with lots of people coming and going, including suppliers for various programs. For maximum secrecy, therefore, much of the development work was done during off hours and under the cover of night. The Blackjack also spent time at GM's proving grounds in Yuma, Arizona, and at both locations it had its own closed-off location within the vehicle development garages. The car was stored and worked on in a separate area that was access-controlled. It never left GM property.

At Milford, another issue was that spy photographers had been taking to the sky to try and snap pictures of upcoming products—and, of course, a new Corvette would be the ultimate prize. For that reason, the Blackjack also drove around with a large, custom-fitted tarp stuffed between the seats. At the first sound of a helicopter or a small plane overhead, the engineers would bust out the tarp and cover the Blackjack, a procedure that Petrucci claims extended by a year the amount of time before the ute-based prototype was first snapped by spy photographers.

Without Blackjack and the critical role it played in validating the mid-engine configuration, the 2020 Corvette Stingray and future variants would not have been possible.

THE END OF THE ROAD

The team refers to the Blackjack as "an engineering proof-of-concept vehicle," and, in time, development work on the project progressed to the point where more production-correct components needed to be tested. For that phase, engineers worked with what are referred to as "architectural mules." Those vehicles had the correct body structure, the LT2 engine, the Tremec DCT transmission, the new car's brake system, the finalized suspension hardware and the new electrical modules. Eleven of those architectural mules were built, five of which were crash-tested.

From there, the third stage of prototyping required over 100 examples. "That stage is effectively where we take that pile of parts that we've designed and stuck together, and we turn that into a Corvette," Petrucci observes. That includes working on

build-ability for the assembly-line workers in Bowling Green, as well as tuning all the components to get the feel that the team wanted for both the Stingray and the Z51.

As more traditional prototypes were built, the Blackjack was ushered into a well-earned retirement. Normally, test mules are unceremoniously scrapped at the end of their useful life. Petrucci and his team, however, knew they were working on a historic project, and they diverted from the standard protocol and recorded their efforts. Part of that was saving for posterity examples of each of the two latter development prototypes and, naturally, the Blackjack. This highly unusual proof-of-concept vehicle in a pickup-truck clown suit would forever be a unique testament to the ingenuity that greeted the challenge of creating a once-in-a-lifetime new Corvette.

Blackjack paved the way for later iterations of pre-production test vehicles and the production version of the 2020 Corvette Stingray (pictured, upper right).

THE DESIGN BRIEF

Designing a mid-engine sports car was the easy part. The hard part was designing a mid-engine sports car with Corvette character.

According to Michael Simcoe, General Motors' global design vice president, the mid-engine C8 Corvette's design brief was relatively straightforward. But that didn't mean it was uncomplicated.

"Corvette has always been something that you can drive like you drive any other vehicle, but you can also drive—if you have the capability—really hard," Simcoe says. "It's something that you could take to run to the shops on the weekend because it was practical and a value-based vehicle. But it's also a track weapon. And, obviously, mid-engine really pushes the performance side and the dynamic side even harder."

As with previous models, the new Corvette's design brief required it be a car you can drive every day, but also one you can hammer hard. Take it to the grocery store or to the track—it does it all.

NO COMPROMISE

The new car thus had to embody all these preordained truths and fit into the model's long history, while undergoing perhaps the most radical design transformation in the nameplate's nearly 70-year life. This was a profound task, in part because of Corvette's extensive and consistent history. Corvettes have been manufactured pretty much continuously since 1953, and since 1955, they have always been available with the most potent V8 GM could muster, lodged up front under its shapely hood.

But the shift to creating a mid-engine car, an everyday exotic, took on even greater significance because of what Corvette represents to the organization—and to domestic car manufacturing. Within the context of the Chevrolet brand, within General Motors as a whole and within the American auto industry, Corvette is a flagship, a halo car for the marque and the automaker—and the country. It thus requires employing the most advanced and forward-thinking design and performance technologies that can be mustered, while always remaining true to itself and its roots.

"It's like any storied nameplate that's been successful. It has its own appeal," says Simcoe—at age 60, a 36-year GM Design veteran—in considering the task before him. "And at that level, it is aspirational."

But a Corvette is also a Chevrolet, an American workhorse. It has to remain attainable, something not completely out of reach—even if it's a big reach—for an ordinary buyer. Herein lies Chevrolet's challenge, as well as the opportunity, when creating the latest version of a long-lived icon. Right until the moment the car was revealed on July 18, 2019, rumors swirled among the press corps and the Corvette faithful that this mid-engine architecture would force the car out of its long-term connection with accessibility, pushing it toward a higher price point, perhaps even one in the low- to mid-six-figure range.

But an eternal Corvette hallmark is its performance-to-value proposition, and Simcoe and his team were not about to let that history fall away just because they were, for the first time in a

A mid-engine Corvette is not a new idea, as this sketch shows. GM has toyed with the concept almost since the car was launched in '53.

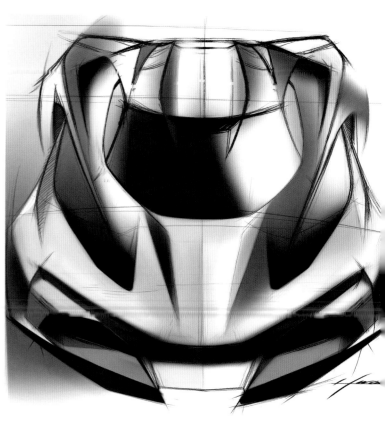

production 'Vette, changing the engine's location from the front of the car to its middle. In fact, this characteristic became a kind of rallying cry for the design team.

"I'll get slammed for this one, but it's very easy to design a mid-engine sports car," Simcoe says. Kids, fans, students in industrial design programs, exotic supercar manufacturers' employees sketch them all the time—typically the more outrageous and extortionate their imagined price, the better. "But to design a mid-engine sports car that exudes Corvette character, that is much more difficult."

Corvette character means a few things, but overall, it is connected to this notion of bringing the experience of automotive exclusivity to a broader audience.

"If your hands weren't tied, it would be very easy to build a mid-engine supercar, like many of the six-figure and higher-priced cars out there," Simcoe says. "It's easier to do it in an unrestricted way, and there's plenty of examples of that. But to deliver value, to deliver the performance, the character and therefore, ultimately, deliver the Corvette, is the trick here."

This raises the question of whether innate compromises had to be made in the ideation of this car—if tradeoffs had to be accomplished to achieve supercar performance at a price point starting below $60K. According to Simcoe, the answer is a resounding no.

"There were some pretty serious negotiations but no compromises—because if compromises were made by manufacturing, or design, or the engineering team or the guys running dynamics, then we'd have ended up with a vehicle that was not balanced. And that presents itself very quickly to both the media and customers," Simcoe says. "The real difficult thing is being able to have the hard conversations and do the balance, to have enough experience in your organization to deliver despite not restrictions, but control."

This is where the team's expertise and collaborative history really came through. Many designers leading the eighth-generation Corvette (C8) team worked together on the previous-generation (C7) Corvette, and some even worked together on the generation before that, the C6. They were fortunate enough to know not only how to collaborate with each other to meet their goals, but also how to work within General Motors in order to get their best ideas through.

"The balance of character, design materials and style are something that the guys in the studios here do really, really well because they design high-performance vehicles," Simcoe says of the team assembled to work on this special and important project. "But they also design vehicles that are bound to deliver what customers need in practical terms, as well."

Knowing these customers is key to providing them with what they need from a design perspective, so this was a key part of Simcoe's brief to the team. "This vehicle is very deliberately high-value for our customer," he says. "But it also is the result of 60-plus years of staying connected to customers and understanding what Corvette customers desire."

In this case, that meant retaining an energetic and assertive exterior design language, wrapped around an interior providing spacious accommodations for two people and their belongings. It meant the highest possible material quality and imbuing it all with cutting-edge technology that offers practical real-world solutions to contemporary needs, as well as those not yet envisioned, while simultaneously showcasing and delivering mind-bending, top-tier performance rivaling vehicles that cost five times as much.

The new car delivers on all this and more. "The base package design of the vehicle obviously is driven by the engine location and the occupants, but also by the level of practicality, with storage space, practical storage space, both behind the engine and in front of the occupants," Simcoe says. "And it has much more premium material usage and design content than the previous Corvettes. So, in this new layout, we've effectively got more dynamic performance, a brilliant new profile to the vehicle, and we have more interior space than the old vehicle, and we have more trunk space." Simcoe adds, almost as if it's an aside, though it obviously is not, "And it accelerates faster."

OPPOSITE: Many of the C8 sketches were done by designers who worked on the previous-generation car and even the one before that. Sketches here by: Brad Kasper, Hwasup Lee, Vlad Kapitonov.

NEXT PAGE: In the end, the C8 is going to be judged based on its ability to continue the nameplate's evolution, among the industry's longest. Corvettes have been of and for their time—pushing the envelope while staying true to their roots.

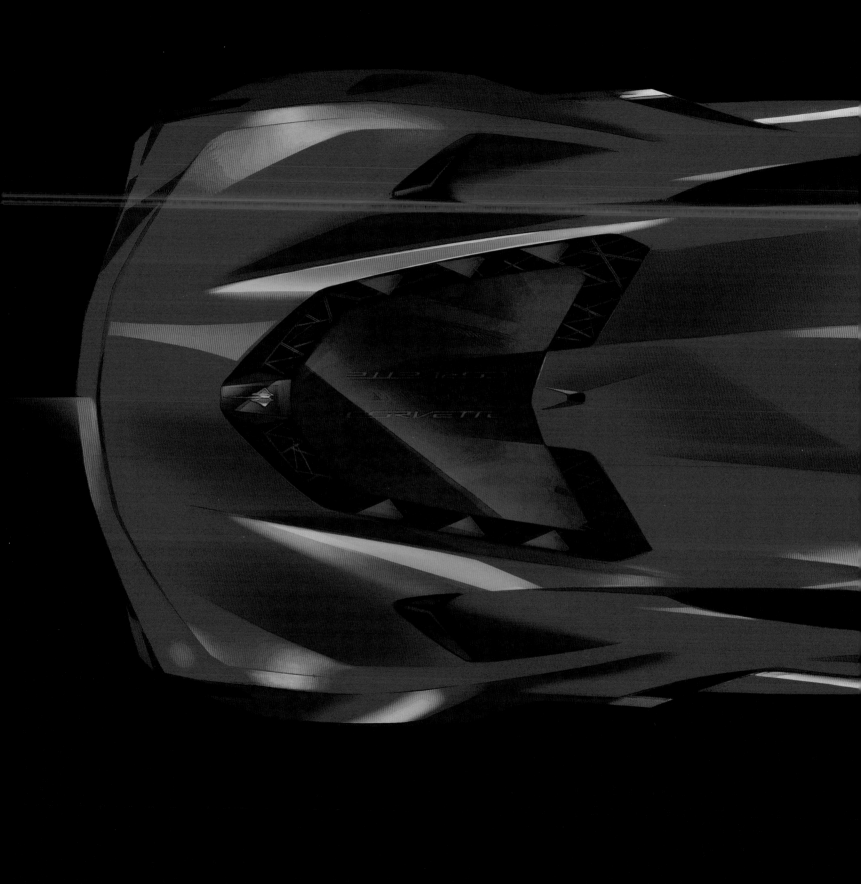

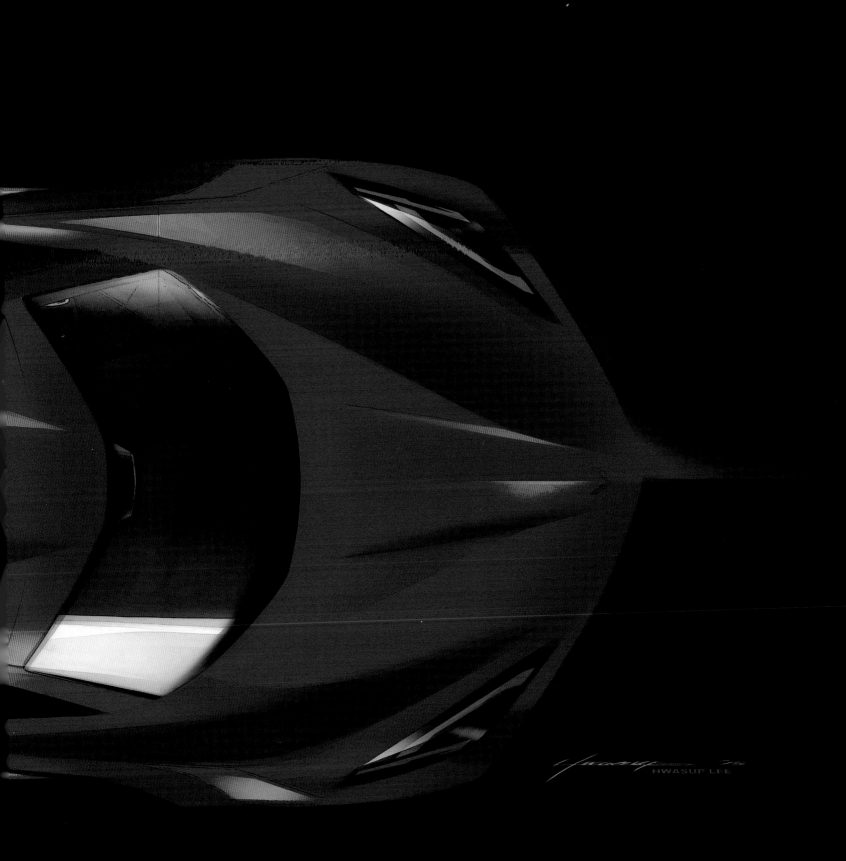

HWASUP LEE

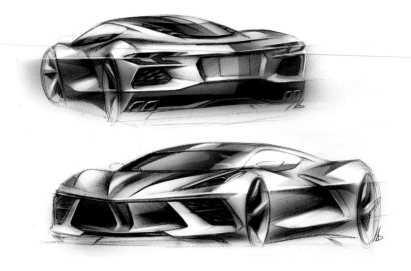

Yet the C8 accomplishes these requirements while still maintaining a visual connection to the nameplate's long heritage, incorporating key cues hearkening back, if not to specific models, then to an overall composite history, a collective and modern Corvette lineage.

"I'd start at the nose and the character of the graphic of the nose," says Simcoe. "Then there's a line, the hard break line, that runs dynamically or diagonally back that defines the body and runs around the rear of the vehicle. There's the surfacing itself, and the character of the surfaces—if you parked a C7 and a C8 together, you'll see what I'm talking about. The character of the graphics, the execution feels the same. And that's not saying the C8 is not done in a different way and more modern, but there's a link."

This connection, especially between the previous-generation car and this one, makes perfect sense. Not just because, as mentioned, these cars were created in the main by the same team of designers and sculptors, but because the C7 was originally envisioned, at least for a period of time, as a modern mid-engine Corvette. "Back as early as about 2007, we had buried in the basement, in

(skunkworks) Studio X, packaging studies for a mid-engine Corvette," says Simcoe of the planning stages for the C6's replacement. "And it wasn't just Design firing a flare to what we'd like to do in the future. It was a genuine packaging study. It was the work we'd done in the wider Corvette organization."

In actuality, a mid-engine layout has been part of the Corvette plan nearly since its inception, as we've seen in other chapters. Experiments with moving the Corvette's engine amidships began under the father of the Corvette, Zora Arkus-Duntov, way back in the mid-1950s, and they have continued pretty much unabated through each successive generation. "It's just a part of the lineage that goes back, an aspiration that goes back 60-plus years," says Simcoe. "Every next-generation Corvette has been a mid-engine Corvette in some respects. If you go back even to C1, it was there. So, certainly, the desire to make a mid-engine vehicle was there from nearly Day One."

So why did it take this long for Corvette to reach what seems like an inevitable eventuality? Simcoe cannot answer this with any certainty. "From a corporate point of view, I can't make a comment on that, simply because I probably can't explain it." But it's certainly not been for lack of trying. "From a design perspective, you've seen the concepts over the years, the ones that have made it out into the public and there's an equal number who've stayed inside," he says. "There's always been a desire from the design perspective to do a mid-engine vehicle just because it's the ultimate expression. You move from the ranks of high-performance sports car to supercar by shifting the architecture."

Ultimately, the C8's success will be judged not on its ability to disrupt convention but on its ability to continue the evolution of this storied nameplate—among the industry's longest, and one that has stood universally for not only being of and for its time, but for pushing the envelope. All while staying true to its roots.

"Many people at the launch who were seeing the vehicle for the first time were surprised," says Simcoe. "During their first look at the vehicle, all they could see was proportional change, obviously driven by the mid-engine configuration. But they all came back within 10 minutes or half an hour of staring at the car, and they said that they could see that it couldn't possibly be anything else." He pauses. "It was unmistakably a Corvette."

OPPOSITE: More C8 renderings. **ABOVE:** The C8 maintains a visual connection to the nameplate's long heritage. The new car is both futuristic and historic at once.

DESIGN HISTORY

While the C8 is the first production Corvette to have its engine amidships, the notion of creating a mid-engine version of America's Sports Car far predates the 2020 model year. In fact, it stretches back almost to the beginning of the Corvette's production, smack in the middle of the 20th century. And while the motivation for moving the Corvette motor amidships was driven mainly by a desire to offer customers greater enhancements in performance, the design of these vehicles is surprisingly relevant and inspirational in the history of the nameplate.

"The idea of a mid-engine Corvette actually predates association with Design," says Christo Datini, who joined GM this year as Manager of Archives & Special Collections and whose team curated an exhibition on the mid-engine Corvette throughout history. "It's really something that Zora (Arkus-)Duntov over at the Chevy Engineering Center pioneered as early as the mid-1950s."

Of course, Arkus-Duntov and his team were not in the practice of fabricating bodies or creating interiors. Once that program settled into its engineering basics, he reached out to GM Styling, as it was called at the time, to assist with what was likely the first mid-engine performance car the automaker had ever created. "The program was actually given a small studio—a semi-clandestine, skunkworks-type studio," says Datini. "I don't know if it was actually called Studio X at the time, as it would be later. And that's where Larry Shinoda created the first mid-engine 'Corvette' body."

Shinoda was an intensely talented designer, who later designed the 1961 Mako Shark concept that presaged the C2 Corvette, as well as the Sting Ray concept that became the C3. He fabricated a body for the engineering concept, an elongated hexagonal Space Age fuselage—molded in fiberglass, of course—with a selfish little single-seat cabin. Aligned with the lightweight chromoly tube frame weighing only 125 pounds, and a 350-pound aluminum-and magnesium-rich engine based off of the 'Vette small block, the body weighed only 80 pounds.

"It wasn't a Corvette, per se," says Datini of the open-seat race car, which came to be known as CERV I, an acronym for Chevrolet Engineering Research Vehicle. "But it's a Corvette in that it was powered by a modified Corvette powertrain."

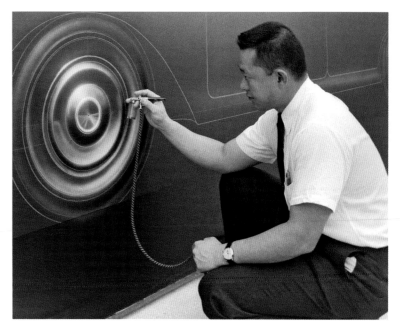

OPPOSITE AND THIS PAGE, TOP: The idea of a mid-engine Corvette started with the CERV I. **ABOVE:** Larry Shinoda works on a rendering in the body development studio. A hot-rodder at heart, Shinoda designed several classic Corvettes.

The CERV I underwent numerous redesigns during its lifespan, mainly to its nose and rear fairing to accommodate cooling or fuel components for the seven different motors it housed during its lifespan as a test bed. But Arkus-Duntov always wanted to race—having had some success at Le Mans in the '50s—so the next iteration, arriving around 1964 and known as CERV II, was designed as a mid-engine racer.

The car included novel features such as variable-split all-wheel drive, but the project itself was stillborn. However, the car's rounded, high-peaked front fenders and flat Kamm-tailed rear, again designed by Shinoda, clearly presaged some details that became key to the sly and curvaceous C3 when it was revealed years later.

Even stronger indications of the C3's future design could be seen in the next Corvette prototype with its engine in an atypical location, the XP-819. This car was developed under a program separate from the Corvette studio, led by Chevrolet Research and Development under Frank Winchell. The XP-819 was actually a rear-engine concept, which appeared in 1964. But it was also styled by Shinoda and definitely demonstrates an overall resemblance in silhouette and form to his Sting Ray concept from later in the decade.

"From what we can tell from the documentation of the programs, the internal design program was actually the same," says Datini. "So, it's completely conceivable that both of the bodies were being designed in the studio, at the same time, as alternatives for a car. It just so happens that there were two cars. But the thing about the XP-819 is that, if you look at it, it's very much like the C3 Corvette. It's a rear-engine car, but it's one of those experimental vehicles that really impacts the production design." It was a progenitor of the C3's high-hipped and high-shouldered "crease-in-the-pants" look.

Other mid-engine Corvette concepts followed, including the 1968 Astro II, again by Shinoda, demonstrating how the C3 shape could be adapted in a more organic way into a mid-engine car with an even lower center of gravity, likely meant to compete with

Larry Shinoda, center, and colleagues look over an XP-819 rendering and foam buck inside Chevrolet No. 3 Studio at GM's Design Center.

the Le Mans-winning GT40, released around that time by Ford. According to Datini, this car appeared as though it could be made production-ready, but it seemed GM executives saw the C3's long-term profitability and decided not to make large platform changes.

"I think that's the big thing about a mid-engine Corvette in this era," says Datini. "You have this C3 Corvette that's doing so well—they're selling so many of them—that Chevrolet management just doesn't want to change. They talk big, but when it comes down to it, why mess with it?"

The mid-engine XP-882, with a transverse-mounted V8 behind the passenger compartment, was the next mid-engine Corvette to appear on the circuit, this time in the spring of 1970 at the New York auto show. From a design perspective, this car broke much new ground, with a squatted-back shape designed by Hank Haga and Jerry Palmer, that was far more angular, wedgy and constrained than the C3. It was almost as if it the existing car had been pulled taut and clipped at the edges. The sharper nose, higher body line and flattened boattail shapes are also clear forerunners of the C4 Corvette that didn't arrive until the mid-'80s.

Now firmly engaged in the oil crisis and shifting emissions standards of the early/mid-'70s, GM attempted to seek solutions to issues concerning fuel economy and exhaust gases by investing in the rights to the lightweight and potentially efficient Wankel rotary engine. This led to developing a series of mid-engine, rotary-powered Corvette concepts: the XP-897, the XP-895 and the aforementioned XP-882. Featuring an array of two or four rotors, these cars made their way through the auto show circuit during the decade.

The standout among them was the final, 1973 XP-882, a four-rotor car with an aerodynamic silhouette. "Bill Mitchell, then head of GM Styling and Design, decided that he wanted it to be teardrop shaped, like one of those Mercedes world speed record-breaker cars in the 1930s," says Datini. "And he sets the design team at that."

The car ended up being stunning, with an overall shape resembling a slightly rounded and flattened isosceles trapezoid, with an extremely low and sharp nose and tail, trifold gullwing doors and a futuristic interior and dash featuring among the first uses of digital instrumentation. It also included a clear display cover for

OPPOSITE: The 1967 Astro II concept car and the 1968 production Corvette coupe. **ABOVE:** Arkus-Duntov with his XP-882 prototype at the 1970 New York auto show.

its signature mechanical component. "The backlight is all glass, so you see the engine quite well," says Datini, noting this element reappeared on the C8.

The mid-engine Corvette project went somewhat dormant during the late '70s and early '80s, with the exception of the Aerovette show car, essentially the XP-882 four-rotor car with a 6.6-liter V8 crammed into in the engine bay. Focus shifted to a replacement for the aging C3. But around the time of the C4's launch in the early '80s, Chevrolet decided it wanted to get back into racing with the Corvette, perhaps to lend further credibility to the C4, which, along with all of the famed '70s sports coupes—the Datsun Z, the Toyota Supra, the Firebird and the Camaro—had grown in scale and weight and developed a bit more of a grand tourer sensibility to it. The result was a mid-engine design called the Corvette GTP.

A Corvette in name alone, it held some vague resemblance in its nose and silhouette to the C4, but was built on a Lola race car chassis, with a body by Randy Wittine. "He had some experience back in the '60s and '70s working on some of these mid-engine cars," says Datini. The GTP had some racing success, but once the C4 was launched, the team began to focus on the next-generation Corvette, and there was yet another decision to try to take the car's architecture to a mid-engine layout.

The next iteration of this design was done by a young designer named Tom Peters, who eventually went on to head up exterior design for the Corvette program. It was known internally as the Corvette MD, but when it was unveiled at the Detroit auto show in 1986, it was named the Corvette Indy. Like many of the era's advanced vehicles, which took advantage of new computer-aided aerodynamic capabilities, the car was rounded, amorphous and windswept. "It had this really wild, organic theme," says Datini.

The concept was meant to showcase technological sophistication, with a small-displacement, high-output, twin-turbocharged engine, as well as four-wheel drive, four-wheel steering, anti-lock brakes and in-dash video screens. Its glaring exposed headlamps also broke with long Corvette convention. From a design perspective, it presaged the C6's rounded shape and double-pin projector lamps, which arrived nearly 20 years later.

Though the Indy was a non-running show car, it inspired a

OPPOSITE: The 4-Rotor concept, circa 1973. ABOVE: The 1986 Corvette Indy with designers Roger Hughet, left, and Randy Wittine, far right.

NEXT PAGE: The 1986 Chevrolet Corvette Indy and the 1990 Chevrolet Corvette CERV III were two cars that lent many design cues to the sixth- and seventh-generation Corvettes.

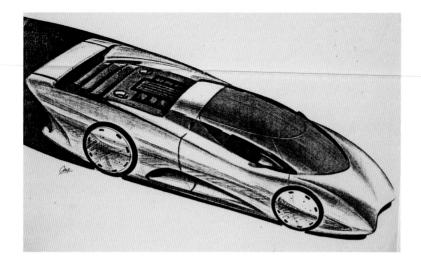

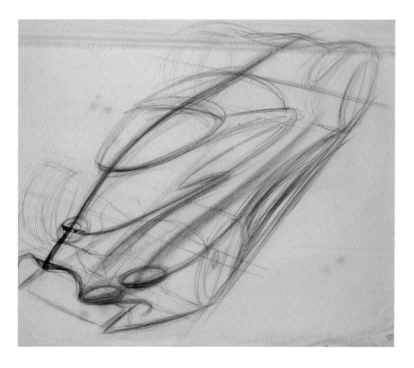

ABOVE, TOP TO BOTTOM: Early sketches of the CERV III and C5 production car. OPPOSITE: GM compared Corvettes and Ferraris on the Design Building patio—among them, a 2-Rotor concept, left; a mid-engine design proposal, center; a 1971 Sting Ray, second from right; and a Ferrari.

fully engineered iteration, called CERV III, which emerged in 1990, featuring a similar silhouette and many of the same technologies. CERV III had sharper lines, with a body made of composite materials including carbon fiber and Kevlar, with aluminum honeycomb reinforcements. Its rounded and narrowed front, thinly coved flank, and flaring and tucked rear clearly demonstrated design influences that later appeared on the C5.

The mid-engine idea, and the whole Corvette enterprise, went through a bit of a rough time in the 1990s. "I don't know that there was a mid-engine Corvette concept after C5," says Datini. "But from what I've read, the idea of a mid-engine Corvette didn't really come back up again until the mid-2000s. This would have been after C6 launched, so in preparation for C7."

Lead Corvette Designer Peters and Design VP Simcoe confirmed there was an entire special skunkworks program centered on creating a mid-engine version of the car that became the C7, but it was scrapped due to the corporate bankruptcy in the late 2000s. Yet the idea of a mid-engine Corvette design never really disappeared. It was endemic to the Corvette's DNA almost from Day One and an intrinsic influence on, and aspiration for, Design. "The designers were totally enamored with that mid-engine idea for Corvette nearly from the start," Datini says. "Bill Mitchell brought in Ferrari Dinos to take a look at, and I want to say a Porsche 914 and a Lamborghini Miura. We have photographs of them on the viewing patio at the Design Center. They're looking at these cars and they're trying to see where they can do better."

This inevitability finally came to fruition with the new C8. "The way Tom Peters tells it, the reason they get back into the mid-engine idea in the present day was because (Corvette Chief Engineer) Tadge (Juechter), he realizes that they can't really go any further with performance without changing the architecture," Datini says.

"And it's interesting because we have all of Zora's papers in the GM archive, and Duntov said the same thing back in the '60s. Even as late as the '90s he was writing to Jim Perkins, then Vice President and General Manager of Chevrolet, saying the same thing. 'You can't go any further with front engine.' He was writing about the Corvette Indy, and he said, 'You have this. You need to make that car.'"

Now, they finally have.

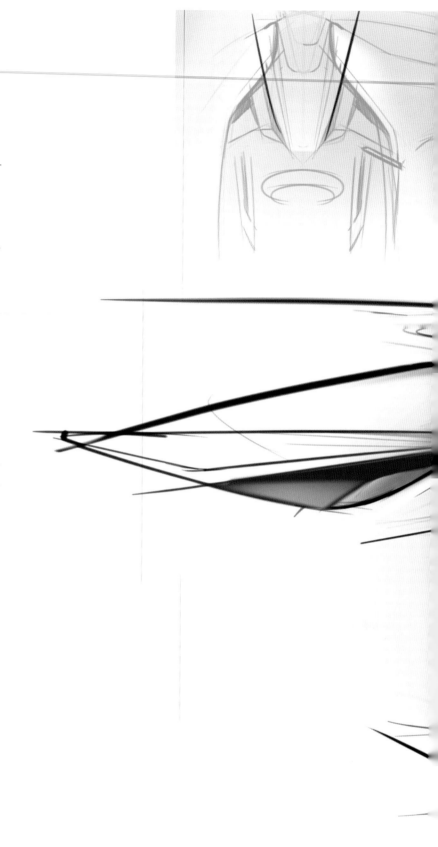

THE INTERIOR

Though its mid-engine architecture required a wholesale reimagining of the car's overall forms, the eighth-generation Corvette was not only being redesigned from the outside in—it was being completely reenvisioned from the inside out. "It was not lost on any of us, or anybody on the design team throughout the process, the monumental task that was set before us," says 36-year-old Tristan Murphy, interior design manager on the Corvette and a 13-year GM veteran.

The challenge of the C8's interior design was compounded by several factors, but an undeniable one top of mind for the team was the nameplate's enduring reputation in terms of the quality and execution of materials, trim and components. This had been significantly improved upon throughout the model's long history and was a particular focus in the creation and run of the new car's immediate predecessor, the C7. Still, the goal of Murphy and his team was "not only to meet the expectations of what the mid-engine car could be, but go beyond the expectations of what a Corvette could be, especially in this price point."

In order to accomplish this goal, Murphy and other members of the interior team sought inspiration at the highest echelons of the car-making industry. Among the most memorable of these excursions was a visit to the exotic-car collection owned by notable Michigan-based collector Ken Lingenfelter. Lingenfelter let the design team loose in his warehouse filled with Bugattis, Lamborghinis and Ferraris. They walked around the room for hours, getting in and out of these exotic supercars, many with seven-figure price tags. They were searching not just for shapes and forms that capitalized intriguingly or effectively on the mid-engine architecture with which they were working, but perhaps more importantly for, as Murphy says, "the experiences that made them feel different from other cars. Even the difference between a more 'high-volume' Ferrari versus something like a very limited-edition car like a LaFerrari."

Some points of inspiration revolved around material quality and

Interior Design Manager Tristan Murphy says the C8 was reenvisioned from the inside out. That's one of his sketches, opposite.

tristan murphy

Designers took a holistic approach to the interior—noting how light and space, and an absence of components, emphasized a sense of rigor and purposefulness. Behind the wheel: Ryan Vaughan, Tristan Murphy.

execution—the way leather is cut and sewn, the way switchgear clicks through its detents. But more important than even this was the team's immersion in a feeling, a sensibility.

The designers noted the way in which light and space, the absence of components, were used to emphasize a sense of rigor and purposefulness. They connected with how a cockpit was selfishly driver-centric and enveloping, protective but also enabling, like bionic armor. And they were delighted to be able to analyze not just what was present, but how it got there. "It wasn't just an inspiration trip to see what materials and forms other exotics used," says 34-year-old Brett Golliff, an eight-year Corvette veteran and currently Chevrolet's manager of color and trim. "It was, 'Oh, this is how they created this.'"

Wanting to expand their horizons beyond the automobile, the designers looked to numerous other areas and industries for inspiration, as well. They analyzed contemporary fashion to inform color and material choices, ones that assisted, according to Golliff, the team's ability to "build stories around emotion" with evocatively hued seatbelts and two-tone leathers. They looked at upscale handbags and luggage for details about leather joinery. "When you make a high-end bag, the seams are decorative," says Murphy, "but they also play up the parts that are functional, letting some of that real-world craftsmanship come into the individual parts." This led to new techniques being applied to exposed stitching on the dash, seats and steering wheel. They looked to construction engineers for lessons in building up layers of massing without adding visual or physical heaviness. "If you look at architecture, it's often these cantilevered elements that can be used to convey lightness or light weight," Murphy says. This led to stretching lines and openings in the dashboard and elsewhere to evoke a sense of speed and litheness.

They even looked to some industries that might seem surprising. "I come from the footwear industry before I came to GM," says Golliff. "For me, that translated very well to Corvette. We just took the motor and made that the foot, and then stretched everything else around it." This immersion led to using performance textiles, adding durability and grip, and reduced weight.

One key step in achieving the overall goal of elevating the interior experience involved the early decision to banish hard plastics as much as possible. The team made this decision before they even started putting pen to paper. "One of the major things that you'll notice when you get into the car is that there's barely any plastic, and if there is, it's in a very hidden area," says Golliff. "And it was very heavily digested as to why it had to be plastic. We really focused on bringing every element up to the echelon that it deserved to be at, in our opinion."

According to Murphy, this advance banishment of plastic was necessary because, otherwise, "when you're designing, you get to afterthought components like the door pull," and the piece gets "costed out" and made of a less premium material. "Even in Ferraris, they'll be painted plastic," Murphy says. "But in the C8, all sides of it are made of premium materials. Two sides are leather or suede, and the other sides are carbon fiber or aluminum. So everything you're touching is premium material." This same premium touch strategy was applied throughout the cabin.

The challenge, of course, was to transform all this varied input into an overall package that was not only cohesive and holistic, but also properly reflective of the nameplate. "As we were developing different design themes, we got pretty far along, and we had started to narrow down the process and we had a design leadership review," says Murphy. "And even though it wasn't a bad review, it was clear that we had some great interior themes, but they weren't uniquely Corvette."

Furthering this process meant not only defining what Corvette is not but, very specifically and organically, what it is—and what C8 would do to further that tradition and move it forward. From an overall perspective, Corvette has, according to Murphy, "always been about making high-end performance at an attainable price." But the new, mid-engine car needed to take that proposition even further. "Not only is it meant to do that with speed or handling, but actually delivering an exotic car experience—the glass over the engine, the glimpse at the proportions over your shoulder when you're at a stoplight," Murphy says. "For a lot of people, that was just a poster-on-the-wall experience. Now,

all of the sudden, it's like, I can afford that." He offers a key caveat. "I mean, $60,000 isn't 'affordable.' But in the scheme of $300,000 cars, this is an attainable car."

The decision was thus made to focus on a design capitalizing on Corvette's very American capacity for democratic accessibility, but to do so in a way that also further enunciated characteristics that had been a part of the nameplate's genetic makeup nearly since its beginning. "We really decided to focus on this very driver-centric, cockpit-style interior, the way the console layout worked," says Murphy. The team's guiding principle was, "What if we just dialed this cockpit theme up, as if we were doing the million-dollar version of that, the hypercar version of it?" The goal thus became finding ways to have the cockpit wrap around the driver, to focus on what Murphy calls the "jet fighter influence," which he believes to be a Corvette hallmark since "at least the C2."

One way the designers sought to accomplish this was to look at the progenitors of this mid-engine layout for Corvette. This meant glancing back in the model's history to the CERV concepts: the single-seat race cars, which were, according to Murphy, "the origin of this whole theme, the cockpit theme."

One big challenge in going with this cockpit theme was how to enact this sensibility without completely alienating the passenger. "Those cars were so nondemocratic. It was like a dictator's view—these are my controls!" jokes Murphy. "But that's part of that whole process, of making it feel like both occupants are involved—that the passenger doesn't feel like they're just over there in their sidecar."

This investigation into Corvette history was a key influence for the designers, not only in terms of what they wanted to do, but also in what they wanted to avoid. "Heritage is a fine line. It can trap you if you allow it," says Golliff. Fortunately, when starting to draft ideas, the designers were pretty much given a clean sheet of paper. "We didn't have this list of what Corvette was or what it was supposed to be—because if we did, that would have designed the car for us in a certain way and felt stiff and not right,"

NEXT SPREAD: A sketch of the eighth-generation Corvette's interior. Note the "jet fighter influence" in the cockpit, which Murphy says has been a hallmark of the car since at least the C2.

OPPOSITE: Another Murphy sketch. He says there was a huge emphasis placed on interior material quality and fit and finish.

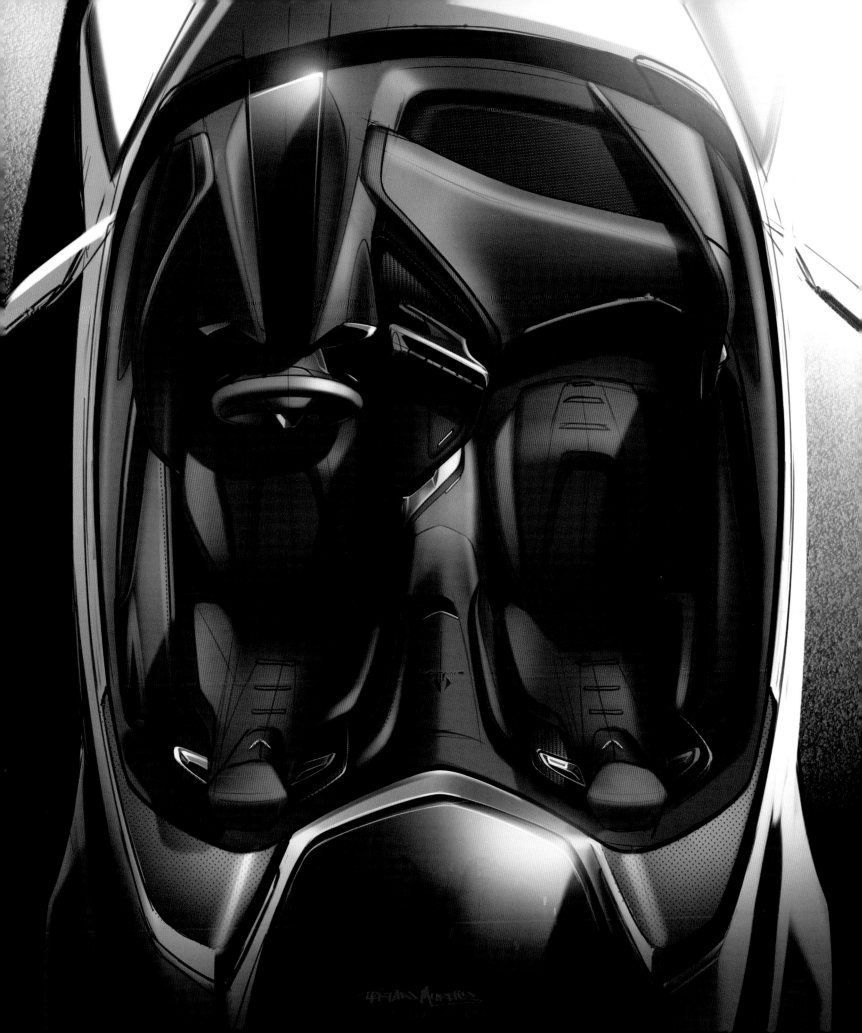

Golliff says. "Where we were lucky was in finding the organic and intangible and making it tangible."

This is not to say they did not include subtle nods to the brand's long history. This was accomplished with colors, like Torch red, long a Corvette signature, or with placement of certain interior components. Take the "waterfall," for example, a body-colored interior trim piece between the seats uniting the outside and inside of the car that features the signature Corvette crossed-flags. This piece has been in existence, a tradition, since the first generation.

In a mid-engine car, the interior layout is a bit different, so the designers couldn't simply return to tradition or update this piece. "So it wasn't about how do we pull back to the original, but how do we keep that element alive, but in a new, completely fresh way that didn't feel like a gimmick," says Murphy. The solution was to incorporate the crossed flags with a subtle perforation pattern in a metal trim piece and then turn the area behind it into functional storage. "So it becomes a nice Easter egg," Murphy says, referring to a kind of insider nod designers often include to delight the faithful. "But one where we're not being quite so ham-fisted with it. It's a more sophisticated way to speak to that."

This follows from the designer's overall theme for the vehicle, one focused on the now, and the future, instead of the past. "Corvette is always pointing forward," says Murphy. "It's always been about trying to make the most sophisticated, progressive, fastest thing we can make."

Eternally driving ideas and practices forward can create some challenges, especially when attempting to achieve as much as C8 set out to—while still meeting its attainable price point. The interior designers faced, and faced down, one challenge after another. "I remember going to Ryan Vaughan, who was the interior director on this project, and we were sitting in a meeting talking about all the different things that we were inventing on the interior," says Murphy. "And I remember telling him in the meeting that, out of all of the programs I'd done, normally you have maybe one or one and a half things where you feel like, if we don't get this to happen, we could be in trouble. With this project, there were so many things that we were inventing for the first time that if we were to have lost one or another of them, it would have torn up the theme. It would have stopped us in our tracks and we could have been in serious trouble from a design standpoint. And I told him, 'I'm kind of nervous.' And he said, 'Good, I am, too.'"

Among the biggest challenges of this forward-thinking, clean-slate approach to interior design was figuring out how to incorporate and take the greatest advantage of the convention-liberating opportunities the mid-engine layout affords. Key among these was excellent forward visibility. "You don't have that big, long hood in front that we've associated with Corvette for so long," says Murphy. "So you can have great down visibility."

In order to leverage benefits from this, to have the most panoramic front and side views, the designers sought to keep the instrument panel and dashboard as low in the car as possible. This meant designing the interior components affecting dash height to be as narrow as possible. The heating/ventilation/air conditioning, or climate controls, vents were key among these parts. Creating an ultra-narrow vent doesn't seem like it should be such a big deal, but in reality, vent size is relatively standard in the auto industry. Suppliers have had lots of practice in figuring how to optimize airflow and air pattern through these conventional grates. So, any change in this metric—especially one cutting the size in half or more, as the C8 team planned—requires some heavy analysis. This is especially true in a car already challenged with air intake and extraction, like a mid-engine car inherently is due to the relatively enclosed nature of the engine compartment.

"Because the opening is so small, we actually had to completely reinvent how a vent articulates, how you move the vanes," says Murphy. "We invented a new way of how air is directed through these things."

Likewise, the designers worked to keep the central touchscreen small, lest they again increase the height of the dash. This meant they couldn't incorporate certain features—like HVAC controls—

TOP: Ryan Vaughan, Tristan Murphy and Brian Stoeckel make sure each and every detail is up to snuff. BOTTOM: The waterfall design, before and after.

into the screen—for optimal functionality, that would ideally require leaving that element of the screen "open" at all times, resulting in a screen size larger than the available real estate would afford. The solution was to create a smaller screen "good enough to do your typical information," as Murphy describes it, but that doesn't obstruct the driver's sightlines or require looking around in order to see down the road. It was also important to angle the screen somewhat toward the driver, but not enough to completely exclude the passenger. "We wanted the driver to feel like it was all about them, but for the passenger to still have a sightline," says Murphy.

This necessitated finding another location for the hard buttons and switches making up the climate controls. Typically located under the central screen, and in two rows, adding an array like this would have necessitated moving everything in the stack farther up, resulting again in a higher dash line. The goal was to figure out how to create hard controls providing tactile feedback and allowing quick inputs without adding to dash height. "And that's where, again, we went back to the jet fighter cockpit inspiration," says Murphy. "Where the controls come down and around you."

The solution was a strip of buttons that envelop the driver's right side in the border between the two seats. (This is the first Corvette to be offered in British-/Australian-style right-hand drive directly from the factory; on RHD cars, it will be on the driver's left.) Again, this lineup had to be low enough so as not to act as a barrier between the two occupants, requiring extensive ergonomic adjustments.

Though it was a major innovation, not before seen on a GM car, what it didn't require, apparently, was a special name. "We just call it Climate Control Strip," says Murphy. His colleague Golliff laughs. "It's one of those things I look back on and think, 'Man, we should have come up with a better nickname for that.' Because we definitely did not have one."

Though it was at heart an engineering-driven decision, perhaps the most controversial and difficult task designers had to con-

From left: Ryan Vaughan, Brian Stoeckel, Nathan Dressman and Tristan Murphy share a moment of levity while poring over yet another critical component of the car's interior layout.

The new steering wheel and the action of the shift paddles had to be perfect, engineers insisted. They wanted a mechanical, high-quality feel to them.

front was doing away with the option for a manual transmission, a feature that had been available in nearly all Corvettes (with exceptions in 1953, 1954 and 1982) since the C1. The manual was given up because the automated dual-clutch transmission, aka DCT, offered superior accelerative, performance and packaging capabilities, and the take rate and certification costs on the standard shift were difficult to account for from a cost-benefit perspective. But it was still not a simple solution.

"We knew from the beginning that we weren't going to have the business case or the space or the design to do both a manual and a DCT," says Murphy. "And that was a huge thing. I know everyone thinks that we just threw the manual out, but that was practically a bigger conversation than going to mid-engine. That was quite a contentious discussion."

The move to a shift-by-wire technology allowed the designers to create a pushbutton transmission selector that could be positioned in whatever real estate was available and convenient, as opposed to the conventional PRNDL shifter location between the seats. This allowed them to move the switchgear far forward and in a narrow band. But it took much trial and error to find a solution that worked comfortably and intuitively. "We were not only doing the design styling and the theming," Murphy says. "We were creating these brand-new inventions."

One interesting ancillary outcome of the move to a single transmission choice was an increased focus on the steering wheel, especially as it pertains to the manual gearshift paddles for the dual-clutch transmission, which are typically located at the 10 and 2 o'clock positions on the wheel.

"With the 2020 Corvette using a DCT exclusively, the feel and action of the shift paddles had to be perfect," says Ryan Vaughan, the interior manager on the C8 project and an 18-year GM veteran. "We wanted them to have a mechanical, high-quality feel to them."

The solution was to make the paddles from cast magnesium and to tune the action of the mechanism extensively to achieve a crisp feel and allow for proper, but not excessive, effort. But accomplishing this, along with discussion of how much functionality should be included in the wheel's face, led to even more scrutiny of the wheel overall. "The first designs we had for the wheel were nice but fairly conventional," Vaughan says. The team was thus

challenged by top executives to "go further and design a wheel that was more exotic and special," according to Vaughan.

The result was a Formula 1 inspired squared wheel, with flattened sections on the top and the bottom. In testing, it was found the top of the wheel could obscure some instruments. Going back to the drawing board, the team discovered if it pulled the corners out further in the flat section at the top, it could eliminate any blockage of the gauge graphics. "This was a big functional benefit," says Vaughan.

Further iterative design—more than 30 different wheels were created, littering the studios—led to the movement of two spokes to 4:30 and 7:30 and the elimination of the traditional 6 o'clock spoke. "It feels great in your hand," says Vaughan, "like gripping a racing wheel."

The final major area to address inside the car was the seats. In the previous-generation car, the C7, there were two seat options: a touring seat, intended for typical spirited driving, and a competition seat, mainly for those who wanted to take the car to the track or use it for more technical experiences and maneuvers. However, this created an issue for many consumers, because the appearance of the competition seat—with its carbon bezels and more expressive sew pattern—was very appealing. "So customers would check that box on the order form," Murphy says. "But then they'd get into the car and find that it was not ideal for everyday purposes, so they found that they wished they'd gotten the touring seat."

Resolving this dichotomy led to what the team calls a three-seat strategy in the new car. "We knew going into it that we wanted a clear delineation between a track seat and a touring seat," says Golliff. "And we knew that we had to find a gap that had the dramatic aesthetic value of the competition seat but the comfort curves of the touring seat."

The result was the so-called GT2 seat, which is visually dynamic and offers excellent grip and hold, but it still provides a level of comfort comparable to the touring, or GT1, seat. (It's the Goldilocks solution: just right.) The team also upped the material quality in this component by offering full napa leather on the GT2 seat, an option also available on the track, or Competition Sport, seat.

Moreover, the addition of the GT2 category allowed the team to increase the capability of the Competition Sport seat, giving

Nathan Dressman and Ryan Vaughan examine the GT2 seat, which was the result of wanting to combine the dramatic aesthetic value of the previous car's competition seat with the comfort curves of the touring seat, Murphy says.

it functionality that skewed even more toward dedicated track use. This had an immediate effect even before the car went on sale, while it was still undergoing in-house tuning. "In the past, usually our test drivers were replacing even the competition seat with Sparco racing seats for testing," says Golliff. "Whereas now, they're maintaining our seat. It's living up to the exact capability that we wanted it to, and that's an incredible feeling."

The quantity and scale of new innovations in the C8 interior is staggering. The team was fortunate enough to be given leeway to dream up unconventional solutions, ones different from those ever implemented in a GM vehicle. But this process was expedited in no small part because they were also granted the capacity to create interesting modifications in the way they worked together.

Much of this derived from the aforementioned familiarity that interior team members had with one another, having worked together on previous projects, including Corvettes. But there was also an unforeseen structural benefit that developed during the design process, an internal and collective sensibility, fostered in part by the confidential nature of the program.

"I've been here for almost a decade now, and I think one thing that was special, that I haven't seen was that we were locked in one room together," says Golliff. "Well, everybody else was locked

out. So we had that element of being secluded and the element of the entire team staying together for the whole time. And because we were all in one area, we weren't waiting around for reviews. It was a constantly evolving and growing design because we were always looking and talking and seeing."

In part because of the close and hermetic nature of the entire Corvette development ecosystem, integrated cross-collaboration between divisions became a more holistic part of the design process. "We were coming to the table fully engaged, not just as stylists making something pretty," says Murphy. "But having thought about why we need it, how we can do it, how we can get it done, how it benefits us from a manufacturing standpoint, and even making this thing run through the cost gauntlet, here's how we can actually still afford it."

Golliff builds on this same idea. "It made it an incredibly efficient process too, because we kind of came into meetings knowing that we'd already solved the roadblock before the roadblock even came at us."

Within the team, much of the credit for this innovative and collaborative spirit is given to Juechter, Corvette's executive chief engineer and the leader of the C8 project. "The most important person in this whole thing, who enabled the trust, who allowed us to do our job, was Tadge," says Murphy. "His goal, his whole focus is what can be done to make the car the best that it can be."

Certainly the fanfare surrounding the release of the C8 speaks to the ways in which this process had an intensely positive outcome. This was highly encouraging to the team, not only because of the way it seemed to indicate something about the future success of this car, but because it gives the team further credibility in helping adapt and advance Corvette to meet ensuing scenarios, anticipated or unforeseen.

"When you deliver on these promises," Golliff says, "when you go beyond expectations of delivering what would seemingly be unattainable things, I think that opens the door to more possibilities—of what else Corvette can be in the future."

ABOVE: Interior team members proudly pose around their work. **OPPOSITE:** Designers put their heads together one more time.

The interior takes the C7 theme to new heights: The C7 hinted at separating the driver from the passenger with a small partition, but the 2020 car has a much larger divider.

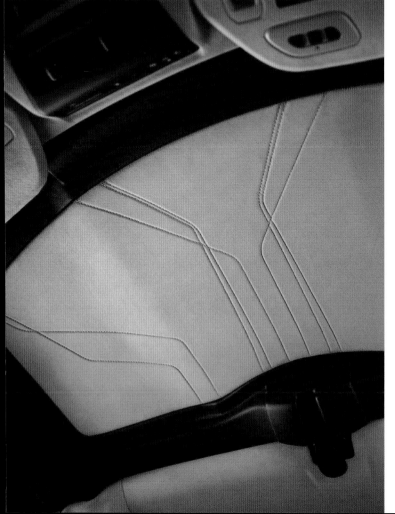

THE EXTERIOR

"The shape, the form, the sculpture has to be the brand," says Peters, the retired director of exterior design at GM's performance studio and a 40-year veteran of General Motors.

This was the first point in the exterior design brief Peters gave his team for the mid-engine 2020 Corvette, a program he led. Peters knows of what he speaks. He has lived and breathed Corvette for decades.

"Some of them can be radically different," he says of the shape, form and sculpture of various Corvettes through the ages. "Look at the difference between the first- (C1) and second-generation (C2) cars. If you're going to make a difference, it has to be profound, and people will grow into it."

Peters recognized the monumental nature of the task before him, shifting the architecture and some of the foundational ideas behind a storied, nearly 70-year-old nameplate. But he wasn't cowed. "You always hear, 'Oh, if it's mid-engine, it can't be a Corvette,'" Peters says. "But that didn't worry me in the least. Because I know that way before I ever got here, and during my tenure here, we worked on several mid-engine concepts and even some race cars, and they all looked like Corvette." (Peters, remember, designed the mid-engine Corvette Indy concept of the mid-'80s.)

That brought Peters to the second point in his brief. "Don't design this for a traditional Corvette person," he says. "We are so close to our customers—we love them and we know how to make a wonderful car that will win their love immediately or they will grow into." Instead, he gave his team some counterintuitive, but thought-provoking, parameters. "Design for a 10-year-old kid. Everybody remembers when they saw their very first Corvette. It needs to just captivate you, like a spaceship that landed. I want it to stay with you for your whole life."

This led to the third briefing point for this genre-jumping, convention-smashing design. "You do Corvette, but you've got to do it in a new way," he says. Moving into this realm, that of the mid-engine exotic car, he wanted the design to immediate-

Everyone remembers the first time they saw a Corvette, so designers were told to design the car "for a 10-year-old kid."

HWASUP LEE

ly communicate its readiness—that it belonged in this exclusive echelon of the automotive industry, one ordinarily reserved for six-figure exotics.

"We were going to the baccarat table and sitting down with Frank Sinatra and Sammy Davis Jr. and those guys now," Peters says. "So we'd better bring the chips."

His team was more than happy to follow and interpret his directions. "For me, the brief was: speed, fluidity, assertiveness, athleticism—and make it look like a Corvette," says Vlad Kapitonov, a 19-year veteran of GM and a key performance car exterior designer. "We were told, 'Just draw these emotions, and we'll take the ones that look like Corvettes.'" The team was told to strip itself of preconceptions and go for the passion points. "Just loud music on, let it flow," says Kapitonov. "That's it."

The goal was to create the pinnacle of attainable American performance. "It was our chance to compete on the world stage with what we think we can do best in an American sports car," says Kirk Bennion, a 35-year GM veteran and Corvette's exterior design manager.

Kapitonov echoes this sentiment. "Corvette is this heroic proportions, super-appealing, just this thing you can't not fall in love with," he says. "There's nothing snooty about it. It's something that's so accessible and amazing. It's kind of part of the American dream."

In order to accomplish this, the team reached immediately into its vast and accessible storehouses of inspiration. They thought first of speed. "We're fans of fast things, like race cars, airplanes," says Kapitonov. "So after I got the brief, I didn't get on the internet and start looking up fighter jets, for instance. That whole attitude is already kind of inside, so I just had to switch it on and do it."

The team created inspiration boards including a diverse array of sources: motorcycles, animals, sports equipment, science-fiction fantasias, Marvel superhero movies, art, architecture, fashion, military hardware, rockets. "These are leading-edge technologies—

purpose-built, serious high-performance machines," says Peters. "And when I say serious, when you're involved with it, you've got to pay attention because it is a huge responsibility, that kind of capability. So you want that to come across when you look at the vehicle, not just when you drive it. It's got to look the part. It's got to have that character and that attitude."

The designers began sketching. And sketching. And sketching. "There's a funnel, basically," says Kapitonov. "It's hundreds if not thousands of sketches from tens of designers." These eventually get voted up or down, and the team starts to produce physical models. This comes in part from Peters' belief in the necessity of the corporeal. "If a picture is worth a thousand words," Peters says, "a model is worth a thousand pictures."

Another thing Peters insisted on in winnowing down the design was the necessity of maintaining a human touch in this process, connecting to the experiential and the emotional. "We have some great digital sculptors that have passion and convey it through that medium, but one thing I was really insistent on was that we had to have hand-sculpting," says Peters. "Because I think it adds soul or personality or a human element to the surface development, to the sculpture. It takes an outstanding car and makes it just superb."

The team made 12 to 15 scale models and then narrowed that group down to three morphing and adaptable full-size models. In considering these, the team and GM executives eventually agreed on two designs. "Mom and Dad, basically," says Kapitonov. "And these got—it's easy to say—blended. But it went on for a couple of months where they existed concurrently. It was, like, you put this detail on that car, and we reinterpret it."

Eventually, it became one car.

Of course, the team faced intense challenges along the way, working up to the eventual codification into one design concept. The biggest issue among all of these was rather obvious right from the start: the architecture of the car. Though mid-engine variants of the Corvette had been attempted in concepts and

NEXT SPREAD: The team faced intense challenges along the way in trying to codify a variety of elements into one design concept, with the new mid-engine architecture being the chief issue.

OPPOSITE: The goal, GM says, was to create the "pinnacle of attainable American performance," to design a car that could compete on the world stage.

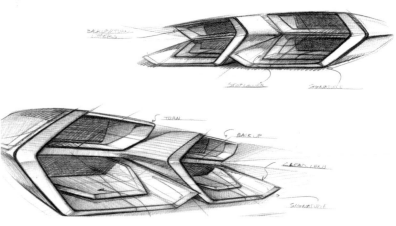

production-prepared variants throughout the model's long history, one had never actually been greenlighted and productionized.

"Within the mid-engine architecture, there's an aesthetic balance, of course," says Bennion. "How far do you move the occupant forward? What generates good side-view profile? The windshield touchdown, the forward accent of the car, the width, the downward vision—but the biggest challenge was getting all the air into this car."

The team had moved from the C7, with a single radiator in front, to the C8 concept. It had two radiators near the corners with condensers, which had brake cooling, which, in turn, had a vented wheel liner. All of these things require air, often in massive quantities, and air requires direction, guidance, harvesting. And control. Thus, the team had to work to design strakes, scoops and spoilers, ones that could work in concert with the bodywork to move the air around and put it in the right place.

Of particular challenge was getting the front corners right. The airflow here was fierce, and it was of crucial importance to keep air attached along the sides of the car and force it into the engine bay for combustion, for cooling and for flow. All of this was even more important, given considerations of future products to be developed off of this same platform. "When you're establishing this architecture, you want it to be one that you can modify," says Bennion. "It's just not about one model."

The resultant shapes—the lines flowing in and out of the bodywork openings, around the engine bay and through the rear—not only act as guides for the almighty air, they also echo key design cues long associated with Corvette. The car's signature "coves" along the side become more pronounced and take on a broadly muscular three-dimensionality as they open to inhale. (They also hide the hidden door release.) The flattened boattail reminiscent of every design since the C2 is not just there for sporty style or aerodynamics—it pushes air in and out of the engine compartment, flowing under the lifted and louvered glass covering the growling V8. The sharp yet rounded nose slices through the air, its prow line flowing all the way back, like a near-invisible homage to the 1963 Split-Window coupe.

Another key challenge revolved around packaging requirements. Though it will have the power, handling and looks to compete with

Exterior designer Hwasup Lee (opposite) and Vlad Kapitonov (above) examine the shapes that came out of the new architecture—which not only act as guides for airflow, they also echo key design cues long associated with Corvette.

six-figure, mid-engine exotics from Italy, a Corvette always has to be an accessible daily driver. This means, in part, that it requires the capacity for an ordinary person to get in and out of it without resorting to tricks ordinarily performed by a Cirque du Soleil contortionist. The narrow sills make this possible. It also must have a trunk large enough to accommodate two pieces of luggage or golf clubs (the removable roof panel must also have room to be stored here, inside the car), as well as additional storage in the frunk—or front trunk—despite the aforementioned airflow challenges in this piece of real estate.

"In order to make it shapely, for the whole rear of the car to stay athletic and beautiful, it's not easy to wrap around a volume like that, a volume that other competitors don't have to deal with," says Kapitonov. "We have to be a lot more creative to make the car look the way it does, for it to stand proudly next to a Ferrari or Lamborghini."

The trunk area thus became something of a "crossroads," according to Peters. "You've got the engine, transmission, suspension, tires, air exiting, hatch latches, tail lamps, exhaust, mufflers, heat, venting, critical aerodynamics where the air departs from that area," he says. "As well as mounting areas for wings and spoilers, and access to storage—not to mention the bumpers and crash standards, the way energy is transmitted through the structure."

Certainly a lot is happening back there. But it all manages to cohere, in part because of the way the rear end is divided into purposeful and horizontal sections: spoiler, tail lamps, heat exchangers and license plate recess, bumper, exhaust.

The midsection of the car was another challenging area, particularly from a packaging perspective. "When you do concept cars, you can do anything you want," says Peters. "But when you get into production, this area becomes the most difficult part of the car."

Not only is it the thickest part of the vehicle, it's the most complex. So much is there: glass, steel, body material, doors, latches, hinges, safety, air intake and egress, and human intake and egress.

NEXT SPREAD: The intense challenge of incorporating the long history of a storied nameplate into a layout completely different from any production Corvette ever resulted in this literal work of art, as examined by Kirk Bennion.

THIS SPREAD: Designers and sculptors constantly worked on the front, midsection and rear of the new Corvette to bring everything together cohesively.

SC-7

JOHN FORCE RACING CAMARO

"It's a very challenging piece of real estate there," says Peters. "And just from a visual proportions standpoint. We would stand outside on the viewing patio looking at models, and I'd just go, 'Damn. If we get that, we'll be good.'" The solution came in keeping the surfaces at once clean and sculptural, lined with cohesive creases connecting to the rest of the body, drawing the eye back along the flanks and through the bulging rear fenders.

One of the most significant struggles revolved around the challenges of incorporating the long history of such a storied nameplate into a layout completely different from any car in the Corvette production pantheon.

Corvettes have always had a sort of bubble canopy on top of a fuselage appearance, along with powerful fender shapes with a creased edge—what Peters refers to as "the crease in the pants." This shorter hood, longer deck and broader proportions complicated this.

The solution was to pull back, not to get lost in layering on details, but to realize the familial relationship between all of the different elements.

"We all have two eyes, a nose, mouth, but there's differences about them that makes your family your family and my family, or anybody else's family, their family," says Peters. "It sounds hokey, but that's kind of it because that gets to my other design philosophy. Yes, it has to have an aggressive attitude, but it has to have a personality to it. It has to have a face."

Kapitonov agrees, claiming that, as a designer, he never wants to ignore the heritage of a model—especially a long-running one. But at the same time, he celebrated the relief of finally getting to be part of the team that brought the mid-engine Corvette to life.

"The historical part goes to the idea of, don't repeat what has been done before, and don't make it look outside of the family. But make it look like it's part of that long family line," he says. "Those past cars are the fathers and the grandfathers. This car, that's a different person. But it's unmistakably a Corvette. It has the Corvette last name."

The key to this integration of history, according to Bennion, comes in ensuring that the Corvette's design remains current. Like all of the best automotive, industrial or architectural design, it must at once define its times, as well as lead them and be fully a part of them. "Corvettes are very reflective of the time, the time they're relevant in," Bennion says. Imagine the thrilling optimism of the immediate post-World War II era as defined by the C1, the Space Age can-do potency as encompassed in the C2, the louche and baroque hedonism of the C3, the accessible technological emergence of the C4 and so on.

The C8 expresses this same finger-on-the-pulse sentiment of our era, a time of rapid advancements and always looking forward. "When we did C6, that was closer to the millennium change (Y2K), and people were a bit more fond of lineage through history," Bennion says. "Whereas now, the lineage thing, everyone's gone through that, so now we're all about what's next."

This shift in cultural perspective, and a consistent and abiding overall mission for the car, allowed potential owners of the new C8—even those among the Corvette faithful—to relate to the design in an interesting way. This, despite the fact that it is markedly disruptive in the overall lineage of the nameplate.

"When we were doing some early studies in consumer research, we would show customers a front-engine car and a mid-engine car, and then we'd show them the performance metrics," Bennion says. "And any time we showed them where the midship would give them a better 0-60 time, we'd ask the focus groups, 'Which car do you want?' They'd all go, 'Give me the fast one.'" This is Corvette. A penchant for speed.

The designers did manage to successfully nail an overall "Corvetteness" in the design without it ever feeling slavishly retro. Kapitonov cites a number of these past-inspired details as among his favorite elements on the new car. He is particularly fond of the main character line, a horizon line nearly encircling the new vehicle, one echoing a similar line on concept Corvettes from 1959 and 1963.

The C8's flattened boattail is reminiscent of every design since the C2 and pushes air in and out of the engine compartment, flowing under the lifted and louvered glass covering the V8.

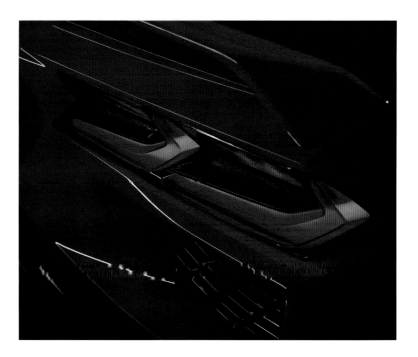

"It functionally splits open for the intake and the wheel. And the fender shapes kind of spill out of it a little, but it's there," Kapitonov says. He claims that in some views, it won't be seen right away. "But a typical view of it will be as it is passing you," he says, smiling. "And that's when you kind of see it best. But it's like a little bit of magic. It's not exactly literal."

Bonnion cites a complementary detail he's particularly fond of: the "leap" to the front and rear fenders. He finds their appearance "muscular," especially given "the low cowl and belt, the break belt sitting in between, which makes the fenders appear even higher," he says. "It's you sitting between the four fender peaks."

Peters reserves his affection for the tail lamps, in part because they demonstrate a different way of interpreting the classic Corvette dual-element rear lighting, from both a technological and a decorative standpoint.

"They use inverse facing reflective LEDs, which we wanted to embrace," he says. "And they have some sculpture in them. It's about composition."

Regardless of individual features, the designers were all extraordinarily proud to have been part of the crew ushering the mid-engine Corvette into existence—an undying dream of the vehicle development team for 60 years, nearly since the genesis of the nameplate, but one that had never before been actualized. Many of them had been through the fits and starts of previous attempts at productionizing the concept, and all along the way questioned whether various moments in the C8's development would be the one in which the plug was once again pulled. But they kept working.

"I think I released my breath only when the cars rolled out on stage in Tustin," says Kapitonov of the reveal event in California in the summer of 2019. "Logically, I knew it was coming, but I only finally felt fully calm and released after they were out."

Though the unveiling was a landmark event, ushering the new mid-engine C8 into actual existence as a production vehicle that consumers could order, it was only one milestone, the start of the latest chapter in the ongoing future of Corvette.

"To me, this is just the beginning," Peters says. "There's electrification, (and) they're developing flying cars. Just think about that. Why can't Corvette be involved in all of that?"

The tail lamps are a modern interpretation of the Corvette's classic dual-element rear lighting. The trunk is large enough to accommodate two pieces of luggage or golf clubs.

THIS SPREAD: Since the C8 is the lowest Corvette ever, there's a hydraulic lift to raise the front end 1.6 inches for going over speed bumps or up driveways.

NEXT PAGE: Some might call this the new Corvette's best angle.

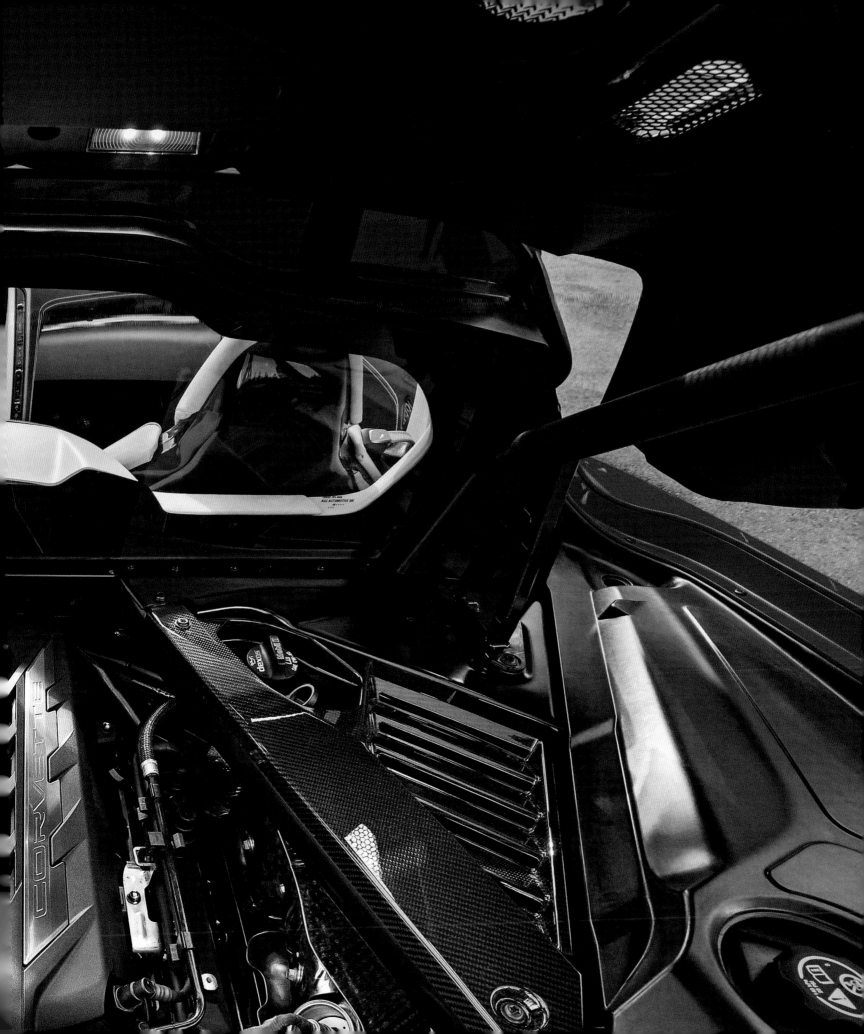

NEXT SPREAD: The much talked about "nacelles" behind the passenger compartment cover the folding top's complex mechanicals.

THIS SPREAD: There was never any doubt the new Corvette would be offered as both a coupe and a convertible. It was planned that way from Day One.

C8 CORVETTE CONVERTIBLE

The mid-engine C8 Corvette was planned from the outset as both a coupe with a removable targa panel and a full convertible. In addition, it was decided early on that this top, unlike those on every previous Corvette roadster since the C1 of the early '50s, would be both rigid and foldable.

"We knew from the beginning that it was going to be a retractable hardtop," says Brett Golliff, Chevrolet's color and trim manager, and a lead designer on the C8 interior. "If you go back and look at all of our sketches, we were designing around the convertible. It was always with the top down, the top off, whatever it may be."

A retractable hardtop was chosen for several reasons, both stylistic and practical. "A hardtop just provides you with a more complete image," says Vlad Kapitonov, Chevrolet performance car exterior designer, who worked closely on the new Corvette. "It gives you a lot more freedom and finishes in the way things are styled. And it's not just like a big canvas canopy—a canopy is just a tent, essentially. It's more substantial."

Kapitonov also notes that, in the exotic-car category C8 is striving to enter, a hardtop convertible is almost de rigueur. "It's more appropriate for a supercar. It's almost hard to imagine going 200 mph with a canvas top—I mean, maybe you could, but you probably wouldn't want to," he says. "This is an aerodynamic, slick canopy that's above you."

The folding composite top's retraction and storage necessitated some significant, but measured, changes in the overall design of the upper portion of the car. The roofline was shifted forward, and the backlight's fastback profile was amended. It also required creating a rounded, conical "nacelle," or fairing, behind each seat—necessary to house the complex five-bar convertible mechanism for the folding top—as well as a new decklid under which the top is stored when it is in its "open" position.

Convertibles are often thought of as slightly less focused when compared to coupes, a bit more of a lifestyle choice than a racetrack weapon, though this new drop-top changes that perception somewhat: It actually sports some additional menace compared to its closed sibling, especially with the top up. "It's almost more overtly mid-engine because the roofline sits even further forward," says Kapitonov. "So it's got a more thrown-forward look to it than the coupe."

The new top comes in three optional finishes. The roof and "nacelles" can both be painted in carbon flash metallic or can both be painted in the same color as the car's body. Or the roof panel can be painted in body color, but the "nacelles" can be painted carbon flash. "The first makes it really sleek; it's kind of doing that black cloth top, but in a much more modern way, to signify it's a convertible," says Kirk Bennion, Corvette's exterior design manager. "The second makes it appear more coupe-like. And the last one

is kind of art nouveau, so it's a contemporary look for how to do a convertible."

Creating all this visual drama required significant strategizing on the designers' part. This was in part due to the general issues associated with creating a convertible: integrating and concealing many more moving parts, as well as fabricating a stylistically cohesive storage area for the top when it is retracted.

For cost and production effectiveness, the team wanted to limit significant changes to the body structure. In particular, it did not want to alter anything behind the engine compartment. The rear trunk occupies this area, with the important storage claim of being able to swallow two golf bags or a weekend's worth of luggage for two. They did not want to diminish this capability.

This left the area between the back of the seats and the front edge of the trunk lid with the task of holding the top and all of its hardware. Unfortunately, in this new model, something rather important already occupies that area: the C8's 6.2-liter, 495-hp naturally aspirated V8 engine. "The main challenge was packaging," says Kapitonov. "It was almost like stuffing 10 pounds of potatoes in a 5-pound bag."

On the coupe, this beautifully rendered motor sits under

louvered glass, amidships and vented to the open air. This placement and style of display not only created issues around storing the retracted top, it also added the need to solve for maintaining proper engine cooling, as well as the means to protect the top from excessive engine heat when folded. Finally, this whole situation called into question how to deliver visual drama when the motor—which, for many on the design team, was the jewel around which the rest of the car was a glorious setting—might be hidden away.

"We needed to make it look just as exotic and interesting and coherent as the coupe, or even more so," Kapitonov says, "while not being able to show the engine because the top kind of folds on top of it."

Further adding to the challenges the designers had to address was that the retracted roof's mechanicals and components added visual and physical height to the area just behind the two seats. "We had to get it all to stack up and still have a decent rearview vision plane," says Bennion. "People have to be able to see out of it."

The solution was to create the aforementioned nacelles behind the passenger compartment. "We had to develop these shapes to cover the mechanicals because there's a lot of mechanization," says Bennion. "When you have the chance to see this convertible operate, and just how the tunnel will open up, and the two half-shells of the top rotate one atop the other and fold, and all that comes together gracefully in half a minute. The mechanization, it's complex in there."

These nacelles successfully help conceal the top when it is down and add some compelling dynamism to the design when it's up. But all of this mechanization operates within very tight tolerances, creating additional design and engineering problems. "There's an interface of interior to exterior here," says Bennion. "So it's one thing to just have the top come down into the tunnel. But then there's an interior part that needs to fold in, fold down and cap the front edge of that in the nacell. And for that to come in as a graceful operation and have all of that interface in that area, it's got to have this airtight fit to show its quality. You need the

Like artists touching up a beautiful canvas, Corvette designers Andrew Casmer (top), Aaron Bunch (opposite left) and Brandon Liscinsky go about the detailed business of perfecting the new C8.

fit and finish, the parts have to align, and it's all got to come together gracefully."

This interior/exterior interplay ended up inspiring the designers, influencing them to consider the nameplate's 65-plus-year history. "The way that early Corvettes were set up, the exterior flowed into the interior, and there was a lot of interaction," says Kapitonov. "With the C8 convertible, this is a lot more overt, especially when the top is down, because the nacelles almost look like they're part of the interior, and they're interacting much more with the exterior of the car than in the coupe."

With the top down, this interaction clearly makes the occupants more visible—and prominent. This is part of a convertible's joy and among the key reasons buyers choose to purchase one. But it also makes the driver and passenger more an integrated element of the car itself. "It's like a roadster that's wrapped around the driver," says Kapitonov. "The interior interacts with the exterior, but it interacts with the driver, as well. The driver is, arguably, almost a part of the design, part of the exterior, almost."

The integrated connection between human and machine was so prominent in conceptualizing the C8 convertible, a human was actually included in the early design studies. "We got this special full-size model and basically cut the top off and started building up things behind the seats," says Kapitonov. "And one of the sculptors sculpted a person, the bust of a person from the chest up, so that would be accounted for in the look of the car."

These exterior elements not only give the convertible a different profile, in certain combinations, they give it a different overall feeling. "The retractable hardtop, our first one, gives the new car a cleaner visual—a more performance and holistic visual," says Golliff. "My favorite variation of it is having the black roof, with the black pillar graphics that come right into the body. It just gives this real sinister and serious look. It's very dramatic."

Typically, a convertible can require significant visual and/or performance compromises over a closed car. But the designers (and engineers) tasked with the C8 worked hard to ensure these are minimal. "The convertible only adds about 70 pounds versus the coupe," says Kapitonov. "So, it's almost nothing. The weight of a child."

OPPOSITE: Designers iron out the smallest of details on the convertible.
ABOVE: Tom Peters, recently retired GM performance director of exterior design, and Kirk Bennion, exterior design manager, sweat the details.

This brings the core difference between the two models down to one thing. "Do you want to have an open and arguably even more exotic profile than the coupe, but don't see the engine? Or do you want to have to remove your own roof and be able to see the engine?" Kapitonov asks. "Both of them will do nearly as well on the track. It just gives the customer two different looks and visual feels, and a little bit different functionality, for no loss in performance."

This, the designers believe, is among the great achievements of this convertible—and the entire C8 enterprise. "That's what's great about the whole Corvette thing," says Kapitonov. "That it gives me something, and it gives somebody else something that is completely not like me." (For the record, Kapitonov says that, personally, he would get the coupe because he would plan to track his car, and on the track, even small differences in weight and aerodynamics make a difference.)

Having these options is wonderful. And it reminds us that all cars have value as specific therapeutic objects. "Convertibles have their own special charm," says Bennion. "You could have your worst day at work, and you can get in one of those cars and put the top down, and by the time you get home, it's like you're worry-free for the evening. It's just the special aspect to a convertible that does that. I can't put my finger on it, but I can tell you that they're a joy to drive."

Having created an individual vehicle that can serve so many purposes is highly rewarding to the designers. But that doesn't mean they are 100 percent satisfied. "In an artist's view, nothing is ever perfect," Kapitonov says. "But this one, I am pretty happy with, overall."

Among the convertible's more impressive aspects? It weighs only about 70 pounds more than the coupe.

NUTS AND BOLTS

Achieving the mid-engine dream took a
ton of work, but it had to be done to take
the Corvette to the next level.

"It's hard to believe

that this is finally happening." That was General Motors President Mark Reuss, on stage in Tustin, California, to introduce to the public the 2020 Chevrolet Corvette Stingray, the first mid-engine Corvette. "Mid-engine has always been part of Corvette's destiny," he said, "and it's something we've been looking at for a very, very long time."

General Motors President Mark Reuss drives the new mid-engine Corvette onto the stage at the car's unveiling in Tustin, California.

THE DREAM IS REAL

The original Chevrolet Corvette, one-time GM Motorama dream car, first rolled into showrooms in 1953, and in the years since, it has represented the ultimate expression of the American sports car. Through seven generations, that vision has been expressed as a front-engine, rear-wheel-drive two-seater. But even as far back as 1959, Zora Arkus-Duntov, father of the Corvette and the model's first chief engineer, began developing mid-engine experimental vehicles in order to explore the outer envelope of performance, in parallel to what was happening in motor racing at the time.

When mid-engine supercars began arriving from Europe in the '60s, the configuration began to be seen as the road to ultimate street-car performance. The idea of a mid-engine Corvette started to take hold—in the automotive media, among the car's fan base and within General Motors itself.

"Even in (Corvette Chief Engineer) Dave Hill's day, we looked at some packaging studies that had a transverse Northstar V8 in the back of a Corvette-like vehicle," says Corvette Executive Chief Engineer Tadge Juechter. "Dave studied that informally to see if it had some merit but ... that was too big a pill to swallow at the time. I don't think he would have gotten a lot of support, just as Zora didn't in his day. So that was quickly scrapped."

The time was not yet right. The dream, however, did not die. It was merely deferred.

Each new-generation Corvette escalated its capabilities. In fact, it was looking more and more that the car might simply run out of runway. Starting arguably with the 638-hp C6 ZR1, Chevrolet was nearing front-engine architecture's practical performance limits. The seventh-generation car was proposed to switch to a mid-engine architecture, but the U.S. financial crisis intervened and the investment could not be justified. Instead, the C7 moved forward with the existing layout. When the car debuted, it managed to push the concept still further, particularly with the wide-body 650-hp Z06 and then the 755-hp ZR1. But as Reuss has noted, "You could make the case that when we got to C7, we had pushed the limits of what we could do in that configuration."

There is also the fact that the Corvette does not exist in a vacuum. In an era when many cars have very high horsepower numbers, it was becoming more difficult to distinguish a Corvette's performance from that of an upper-echelon muscle car or even a high-end sports sedan. What is the reason to purchase a Corvette? The car has to be a performance statement. "To take performance and driving dynamics to the next level for our customers," says Reuss, "we had to move to mid-engine."

Juechter adds: "The bottom line is, we had pretty good technical proof showing that in order to move the car forward from an absolute performance standpoint, we were going to have to do this. It wasn't a question of if, it was a question of when." What was it that enabled that when to be now? C8 Corvette Chief Engineer Ed Piatek points to three things: "I think it's a combination of technology, materials and computer-aided engineering tools."

THE WAY FORWARD

A mid-engine car's advantages are most notable in vehicle dynamics, and they're particularly evident as you reach higher levels of performance. Moving the powertrain, the car's heaviest component, toward the middle of the chassis pays dividends in acceleration, braking and handling. When accelerating, having the engine in the middle puts more weight over the drive wheels, increasing traction and allowing torque to be more easily converted into acceleration. Braking creates a weight transfer that naturally loads the front wheels, so moving the engine mass rearward helps spread the load more equally among all four tire contact patches. This allows the rear brakes to contribute more, shortening stopping distances. In cornering, having the engine weight more toward the middle brings it closer to the center of rotation of a turning vehicle, resulting in less energy needed to execute changes in direction, making for more nimble, responsive handling. The Corvette team sought to maximize all those advantages and more.

OPPOSITE, TOP TO BOTTOM: The first Corvette debuted in 1953 at the Waldorf Astoria in New York. Corvette engineers say they took the front-engine Corvette's performance as far as they could with the C7.

THE RIGHT STUFF

The foundation of any new car is its platform. The 2020 Corvette uses a backbone architecture, as the car has done since 1997. But while a central tunnel is again the chief structural component, it is redesigned for the new mid-engine platform. The front and rear frame elements are attached to the central backbone architecture, and the traditional side rails have been eliminated. In a Corvette first, the new platform also has the flexibility to accommodate right-hand drive, sure to be a boon to Corvette enthusiasts in the U.K., Australia and Japan.

Managing crash energy is much different in a mid-engine design than in a traditional front-engine vehicle. In a mid-engine car, the central tunnel has to provide the chief energy management for crashworthiness. It must prevent the forward movement of the powertrain in a frontal collision, and the energy path of the front structure must feed into it. The distance from the front of the car to the passenger compartment is much shorter than in a front-engine design, and the front wheels are immediately ahead of the pedal box, creating additional challenges. And because the team did not want large doorsills that would impede entry and exit, crash energy had to be directed to the central tunnel rather than to side frame rails.

The vast majority of the C8's structure is aluminum. Fortunately, Corvette's history with aluminum meant the material was familiar to Corvette engineers. The C6 Z06 used an aluminum frame, as did all versions of the C7. The team had learned about aluminum's crash performance, its durability and its fatigue strength, as well as the various joining technologies (welding, bonding and mechanical fastening). All that knowledge could be leveraged for the mid-engine Corvette.

"The fundamental change here was to move away from hydroforming and other types of construction that were primary in previous generations, and go to these high-integrity, high-pressure die castings," says Piatek. They're large pieces, roughly 3 feet long, with convoluted shapes to package around wheels, powertrain components and fuel tanks. Those six beams are the car's bones.

Their casting varies in thickness—some areas are as thin as 3 millimeters and others much thicker. Using advanced tools and optimization, the engineers were able to put the metal exactly where it is needed, so the pieces end up being extraordinarily stiff and extremely strong but also very light. They help make the new car's body structure more than 10 percent stiffer than the outgoing model. If one were to make these parts from separate pieces and fasten them or bond them together, they would weigh more and be less stiff. "The results of the stiffness that comes from this casting are something you can feel when you drive down the road," says Piatek. "It just makes the whole car better."

The supply base able to make these complex castings is limited, especially in the volumes the Corvette requires. "If we were to buy them, they'd be very expensive, and we'd be at the mercy of the few suppliers in the world that are able to make them," Piatek says. So General Motors chose to make them itself, in a powertrain casting facility in Bedford, Indiana. That location gave the components their nickname: the Bedford Six.

SMC, sheet molded compound or fiberglass, is used for the body panels and elsewhere. Not just one type of fiberglass is employed, however. In places where noise abatement is more important, the fiberglass is thicker and is made with a different resin, while in places where that's not important, lighter SMC is used. Carbon fiber is used judiciously, so much so that it appears as a structural element in only two areas: an underbody panel running beneath the central tunnel and the curved rear bumper beam—the latter an automotive first. Magnesium appears in the dashboard and doors. The sophisticated mix of materials results in a dry weight of just 3,366 pounds.

The platform is any new car's foundation: While the 2020 Corvette uses a backbone architecture like every one before it since 1997, the central tunnel has been redesigned for the mid-engine platform.

FROM LT1 TO LT2

The new Corvette's heart is again a naturally aspirated overhead-valve, small-block V8, and in that respect, it is the major mechanical element most closely akin to its C7 counterpart. And yet it, too, is substantially transformed. The new OHV engine is called LT2, an evolution of the previous car's LT1. Elements common between the two include an aluminum block and heads, pushrod-actuated valves and continuously variable valve timing. At 4.065 inches by 3.622 inches, the bore and stroke are the same, and the result is the LT2 displaces 376 cubic inches, or 6.2 liters. Direct injection and a 11.5:1 compression ratio also carry over. The major differences between the LT2 and the LT1 come as a result of the changes in location and lubrication.

"The predominant changes that we made were in the breathing system," says Mike Kociba, assistant chief engineer for small-block engines. The intake manifold and exhaust manifolds were completely redesigned. The engine's move from the front to the middle of the car naturally altered the path for incoming air. A new, shorter intake tract feeds a now-rearward-facing throttle body. The newly designed intake manifold features equal-length runners. Each are now 210 millimeters long, optimizing the flow into each chamber, to capture as much intake charge as possible to produce maximum torque and power.

The new engine location similarly led to newly designed exhaust manifolds. Whereas in a front-mounted engine the exhaust gases are directed down and away toward the underbody, here the reshaped exhaust manifolds curve upward as they route exhaust gases toward the rear of the car. The exhaust manifolds have a four-into-one design tuned to optimize the flow for this engine. Exhaust gases travel from the manifolds to the catalytic converters and the tailpipes, then exit the body through the lower rear fascia via four square exhaust outlets. An optional active exhaust system alters the engine sound using flaps inside the muffler.

For the first time ever, dry-sump lubrication is standard on even the base Corvette. This ensures adequate oil lubrication under the lateral forces the car is able to generate, in excess of 1 g. "The C8 has tremendous track capability," says Jordan Lee, GM's global chief engineer of small-block engines. "Lateral acceleration exceeds the C7; we had to make sure the engine was up to the task."

Previously, a dry-sump system was included in the optional Z51 package, and it used a single pump. The new system has been designed specifically for the LT2, employing three multistage scavenge pumps. The greater effectiveness of those three pumps allows for a shallower oil pan, enabling the engineers to mount the engine more than an inch lower in the body. That helps lower the car's center of gravity.

A new oil cooler has 25 percent more capacity than the previous cooler, and the total oil requirement is down from 9.7 quarts in the LT1 to the LT2's 7.5 quarts. That oil is 0W40 synthetic—a special formula developed with ExxonMobil—and is suitable for track use as well.

"The heart of the engine, the camshaft, was also redesigned," says Kociba. To optimize flow, the team added 18 degrees of exhaust duration and 4 degrees of intake duration. They also added lift to the exhaust, so the LT2 has 14 millimeters of lift on both the exhaust and the intake.

Another change with the LT2 is the coil packs relocated from the top of the valve covers to the side of the engine block. That moves them and the plug wires away from the heat of the now-upward-sweeping exhaust headers. A side benefit of the change is a cleaner look for the valve covers.

The V8 is fed by dual fuel tanks, as has been the case since the Corvette adopted a rear transaxle with the C5 generation. They straddle the engine and are completely enclosed in a metal structure for crashworthiness. Like the LT1, the LT2 features Active Fuel Management, a fuel-saving, cylinder-deactivation feature shutting down one cylinder bank under light engine loads. The switching between four- and eight-cylinder operation triggers a vibration, one easily absorbed by the previous torque-converter automatic.

The new dual-clutch transmission, however, has less inherent capability to mask engine vibration, which is why cylinder-deactivation technology has so rarely been used on a car with a dual-clutch transmission. The C8 engineers were able to marry these two technologies in the new Corvette, however, by putting some micro-slip into the clutch control of the DCT. As a result, those transitions are almost imperceptible, or as Kociba claims "butter-smooth."

ABOVE: Thanks to better breathing, the LT2 maxes out at 495 hp—that's 35 more horses over the LT1.

NEXT PAGE: A true jewel under glass: This is the engine cover for the LT2. Black is standard, but it can also be had in red or silver. Aesthetics matter here.

ALWAYS WELCOME: MORE POWER

"Another thing we wanted to do with the LT2 was increase the power," says Lee. "We added 35 more horsepower beyond the LT1, which was 460. Now we're up to 495 hp." The LT2's extra power came largely from changes in engine breathing, the camshaft and the lubrication system. With the performance exhaust (part of the Z51 package), the LT2 makes 495 hp at 6,450 rpm. The standard version has 490 hp. The LT2 also produces a peak of 470 lb-ft of torque at 5,150 rpm—465 lb-ft with the standard exhaust—and revs to 6,600 rpm. As Piatek says, "There are several places where even our computer models didn't predict how well the car would turn out—one of them was acceleration."

Chevrolet says the Corvette Stingray with the Z51 package accelerates from 0-60 mph in less than 3.0 seconds. That puts the 495-hp Z51 Stingray just a fraction of a second behind the 755-hp 2019 Corvette ZR1, which rockets to 60 mph in 2.8. That juxtaposition provides an indication of what a huge leap in performance the C8 represents—and how much potential still remains in this car.

A JEWEL UNDER GLASS

Engine aesthetics were more important for the LT2 than for any previous 'Vette engine. Working with their colleagues in Design, the small-block team massaged the appearance of engine parts ranging from the engine cover (black is standard, but it can be had in red or silver) to the exhaust manifold heat shields (whose dimpled texture took multiple tries to get right) to items as mundane as the fasteners and their coatings. Tom Peters, a now-retired Corvette chief designer, was a driving force behind the effort. His mantra: "The car is the setting and the engine is the jewel."

That jewel is visible under a 3.2-millimeter-thick glass cover. The powerplant can be further dressed up with the optional Engine Appearance Package, including carbon-fiber accents as well as LED lighting for the engine compartment.

With it, the Corvette's beating heart can be admired even after dark.

KEEPING COOL

Cooling the powertrain is a major engineering challenge in a mid-engine design. For the 2020 Corvette Stingray, primary cooling is achieved with radiators and condensers in the outboard front corners of the body, taking advantage of the Ram Airflow at the front of the car while still leaving room in the center for the front trunk. The side coves feed air into the engine compartment—both for induction into the engine and to cool the engine compartment. For increased track capability, the Z51 adds an additional radiator in the passenger's-side cove, and that radiator works as a sub-cooler for engine and transmission oil. In markets with ultra-high ambient temperatures, such as the Middle East, another supplemental radiator is located in the driver's-side cove. Additionally, the engine and transmission each have their own liquid-to-liquid oil coolers.

There is also venting in the rear hatch so in-city traffic hot air can convect up and out of the engine compartment through a chimney effect. Additionally, a fan powers the airflow through the engine compartment and vents it to the outside.

There is a rear trunk located right behind the engine, and the team worked to make sure that area also was protected from excess heat. "The goal was to make it so that we could put a chocolate bar in that rear trunk and have it not melt," says Kociba. It passed the Hershey test. "We did a trial ourselves and put a chocolate bar back there, and sure enough, it did not melt."

The 6.2-liter, 490-hp V8 looks stunning under the 3.2-millimeter-thick glass engine cover. Opt for the Z51 package and horsepower is 495.

"IT HAD TO SOUND LIKE A CORVETTE"

A small-block V8's melodious burble has been a key part of the Corvette driving experience for more than 60 years. As Lee says, "One thing people really love about the small block is the thunderous sound when you start the engine up. And we wanted to ensure we captured the essence of that character and brought that into the next-generation Corvette." The engineers worked particularly hard on the exhaust manifolds' design to get the exact tones they were looking for. The trouble was, the driver couldn't hear them.

Lee says: "When we ran our first vehicles, those standing outside the car thought it sounded great. They'd watch it go by at wide-open throttle, and it just sounded so beautiful, but it was kind of sterile inside the car. A lot of the sounds that raise the hair on the back of your neck were gone. The NVH engineers had a challenge to try and get the noise back into the passenger compartment."

In a front-engine car, the V8 sounds are transmitted naturally, but that's not the case with mid-engine. And opening up a pathway between the engine bay and the passenger compartment invited in a cacophony of unwanted noise because the accessory drive is located at the front of the engine, and the noises from the accessory drive are the sounds you don't want to hear. With the accessory drive now inches from the driver's ear, the question became: How do you block out that noise and still hear the induction and exhaust sounds?

The solution started with really good insulation between the passengers and the accessory drive. The midglass behind the occupants' heads is laminated two-pane glass like a windshield, except twice as thick at almost 9 millimeters. There are also barrier materials in the bulkhead between the passenger compartment and the engine compartment to further deaden sound.

Once they had effectively blocked the unwanted noise, the engineers had to figure out how to transmit the induction and exhaust sound—the engine music—to the driver. Since the noise can't penetrate the bulkhead, the engineers routed the intake tract in such a way as to bring that sound nearer to the occupants.

The engine air-intake tract stretches from the throttle body atop the engine and curves around to the upper part of the side scoop. That's where the air comes in. As a result, airborne noise is very close to the occupants' ears, and it just has to penetrate the relatively thin door glass. Also, the body structure itself is used as part of the ductwork, creating structure-born noise as well.

"We knew it had to sound like a Corvette," says Juechter. "But we had to think about it in architectural terms to get the noise around that opaque barrier so the driver and passenger can hear it."

A NEW GEARBOX

Supplier Tremec custom-made the M1L eight-speed dual-clutch transmission just for this car. The dual clutch represents another Corvette first; its performance advantages include uninterrupted torque flow during shifting. At the same time, the gearbox is programmed to deliver super-smooth low-speed behavior on par with a conventional torque-converter automatic.

An ultra-low first gear takes advantage of the traction inherent in a car with the new Corvette's rear weight bias. Second through sixth gears are closely spaced, designed to keep the engine in the heart of its powerband. The top two gears are tall cruising ratios for relaxed and economical highway driving.

Inside the gearbox, one shaft has the even-numbered gears (second, fourth, sixth, eighth); another has the odd-numbered gears (first, third, fifth, seventh). The dual-clutch design means the transmission can be simultaneously disengaging a gear on one shaft while engaging a gear on the other shaft, allowing shifts quicker than 100 milliseconds. Torque flows from the gearbox to the rear wheels via a mechanical limited-slip differential. The Z51 package brings an electronic limited-slip differential and a numerically higher final-drive ratio.

Shifting is by wire, so there is no conventional shift lever. A strip of buttons on the center console comprise the Electronic Transmission Range Selector, used for selecting drive, reverse, neutral and park. There are steering-wheel-mounted shift paddles, providing drivers the ability to manually select ratios, and the gearbox features rev-matched downshifts.

Those paddles have one additional function. "We wanted to make sure this car has all the Corvette character," says Kociba. So, when a Stingray driver stops at a light and wants to rev the engine, pulling back and holding both paddles instantly puts the car in neutral so "you can rev the engine to your heart's delight." Release the paddles, the clutches engage, and away you go.

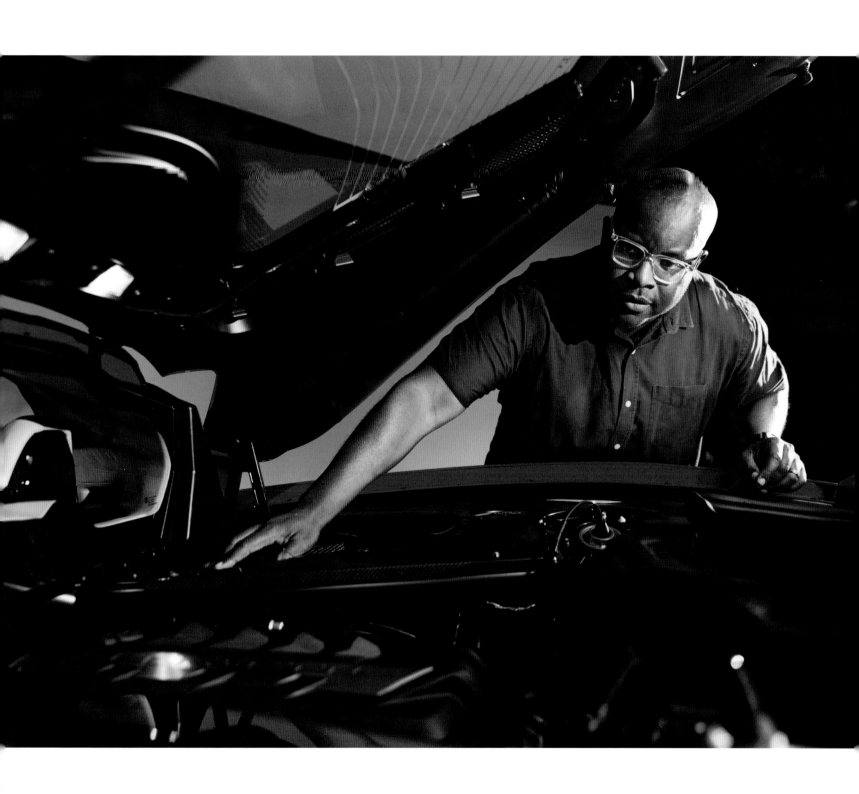

Guy Samuels makes sure everything is in its right place on the 2020 mid-engine Corvette.

CHASSIS CHANGES

Changing the C8's chassis layout afforded the team the opportunity to "reset the clock" as Piatek characterized it, with regards to chassis tuning. One of the most evident changes from past Corvettes is that the C8 uses coil springs over dampers with control arms at all four corners.

Yes, the Corvette's previous and long-serving composite transverse springs are gone. They were unusual, but Juechter calls them "a unique and efficient solution."

With the engine and transmission in the C8 mounted so low in the chassis, however, there was no cross-car path to locate a transverse spring. And once that spring type couldn't be used in the rear, it could not be retained in the front, either, because of the need to match the wheel-rate progression between the front and the rear wheels. So, when the rear wheels switched to coilovers, the fronts had to follow.

In tuning the new setup, the team was not looking merely to eke out fractionally quicker lap times. There was a conscious recognition that customers attracted to this new type of Corvette would have different expectations than many of the car's traditional customers might have had and that the C8 would have to up its game regarding refinement.

In order to appeal to a broad spectrum of buyers, Chevrolet is offering a variety of different levels of suspension firmness and sophistication.

The base package is FE1, and its standard tire fitment is Michelin Pilot Sport All-Season 4ZP run-flat tires, which speaks to the everyday usability the team wanted to engineer into the 2020 Stingray. Next up is the FE3 suspension, bringing stiffer shocks, increased spring rates and thicker antiroll bars. The FE3 setup is part of the Z51 package, which also includes Michelin PS4

performance tires, larger brake rotors with Z51-branded calipers, adjustable threaded spring seats, enhanced powertrain cooling and brake cooling, an electronic limited-slip differential, a shorter final-drive ratio, a performance exhaust, and a Z51-specific front splitter and rear spoiler. It elevates the Corvette "to a racetrack beast," as Lead Development Engineer Mike Petrucci puts it.

The FE4 is the top-tier setup. It includes everything from the FE3 package along with Magnetic Ride Control magnetorheological adaptive dampers, adjusting the damping firmness as quickly as every 10 to 15 milliseconds.

The Corvette's rolling stock aren't just staggered width—at 8.5 inches front and 11.0 inches rear, they're also staggered in diameter, with the fronts at 19 inches and the rears at 20. The wheels are spun-cast aluminum, offered with two different spoke styles and two finishes.

A front-suspension lift system is another Corvette first. In just 2.8 seconds, it raises the car's nose 1.6 inches to clear speed bumps or steep driveways.

It works in response to the driver pushing a button on the center console, but it also can be programmed to remember up to 1,000 locations and store them in its GPS memory, so the driver doesn't have to remember to activate the system every time for a regularly encountered obstacle.

The staggered tire fitment, the tire compound and design, the suspension tuning and geometry all promise to make the C8 more comfortable, nimble and stable than any Corvette before.

"With the new architecture that shifts the engine mass over the driven axle and the move to a coilover suspension, we actually ended up delivering a car that is significantly better for ride and also significantly better for handling," says Piatek. "Usually those two are diametrically opposed."

The team looks over the new suspension setup. Options include the FE1, FE3 or FE4 packages to appeal to a broad spectrum of buyers.

STEER CLEAR

The mid-engine configuration has the driver compartment moved forward in the body, creating a short, straight steering system, one that in the Corvette is 50 percent stiffer than in the outgoing car. That makes for exceptionally direct steering feel. The 2020 Corvette again uses electrically assisted power steering, now with a quicker, 15.7:1 ratio for more immediate response.

With the standard suspension, the Corvette's turning circle is 38.1 feet curb to curb.

But the optional FE4 suspension reduces that by allowing the front wheels to turn more sharply at low speeds when the front suspension is not articulating over a bump—cutting the turning circle to as tight as 36.4 feet.

AS YOU LIKE IT

The performance spectrum of the 2020 Corvette Stingray is broad, and the car allows for greater configurability than ever before through the standard Drive Mode Selector. Four basic programs are preset: They are the familiar tour, sport, track and weather. They're accessible via an ergonomically designed knob prominently located on the center console next to the gear selector. Each setting now customizes more aspects of the vehicle: the steering

effort, quickness of the throttle response, the transmission shift mapping, active fuel management, suspension damping firmness (with FE4 suspension), brake-pedal feel, the engine sound, and the programming of the traction control and stability control. Each mode also alters the appearance and the layout of both the digital instrument cluster and head-up display.

"There's a huge amount of bandwidth between tour, sport and track modes now," says Juechter (above). "The modes are surprisingly different, to the point where you don't just pick one mode and stay with it because that's your favorite. You'll find that people really do want to switch modes based on the road that they're driving and the way that they're driving."

To that end, Corvette engineers have provided drivers with two customizable settings: My Mode and Z Mode. My Mode allows drivers to mix and match from many available parameters to create their own ideal setting. Z Mode does the same, adding in the powertrain parameters.

The idea with Z Mode is that drivers can custom-build their own more sporting setup tailored to a special-use case they regularly encounter—say, one specific stretch of an oft-driven road. With Z Mode accessible via a button on the steering wheel—labeled, simply, "Z"—that special setup is instantly available.

The C8's body structure is 10 percent stiffer than the outgoing model. The Z button, above, accesses a customizable drive mode, adjusting suspension, steering, throttle response and much more.

GETTING AN EBOOST

The Corvette launches a new braking technology with its eBoost brakes. This is a brake-by-wire system providing advantages in mass, packaging and tunability.

A conventional hydraulic brake system has a vacuum booster, a master cylinder and ABS modulator. In the brake-by-wire system, the eBoost module replaces those three components, providing tremendous packaging advantages in the front cowl area and saving weight as well.

From a driver's standpoint, the more notable breakthrough is that the system delivers a consistent brake feel throughout the pads' life and in different temperature conditions.

It also gave the engineers the ability to tune the brake feel. For the C8, there are three different calibrations. The tour calibration has a moderate, progressive feel. Sport increases the braking force through the whole range of pedal travel. The track calibration doesn't merely further amp up response; instead, it changes the shape of the response curve so the initial bite is

comparable to sport but then as the driver goes deeper into the travel, the gain becomes more spread out, providing greater modulation under harder deceleration as one nears the ABS intervention point.

With the C8's rear weight bias, the diameter of the rear brake rotors is greater than the fronts, with both the base and the Z51 equipment. The standard (JL9) vented rotors are 12.6 inches in front and 13.6 inches rear.

The Z51 package's upgraded units (J55) are 13.3 inches front and 13.8 inches rear. Brembo supplies the calipers, pads, and the J55 rotors. In the standard system, the rotors are squeezed by two-piece aluminum four-piston calipers, while the Z51-spec brakes use monoblock four-piston aluminum calipers. The Z51 system also uses different pad material. The calipers are available in four colors: black, yellow, bright red and Edge red.

An electronic parking brake is another Corvette first. It activates a secondary set of rear calipers rather than a drum-in-hat parking brake unit, saving nearly 7 pounds.

The car launches a new braking technology, eBoost brakes, a brake-by-wire system providing advantages in mass, packaging and tunability. The calipers are available in four colors: yellow, Edge red, black and red.

PUTTING IT ALL TOGETHER

Clearly moving the engine to the middle of the car has altered the Corvette's packaging more dramatically than at any point in its history, and one major consequence is the passenger compartment moves forward a whopping 16.5 inches. Combine that with the dramatically altered weight distribution you get from moving the engine just ahead of the rear axle, and the car's center of gravity ends up at the same place in the vehicle as the occupant's center of gravity. "So," Juechter says, "The car literally rotates around the driver in a turn."

The placement of the passenger compartment further forward and the driver no longer having to peer over the engine have both transformed forward visibility thanks to a much lower cowl and instrument panel. There also are subtler aspects aiding outward visibility and packaging overall. "We have tighter clearance between components than any car I've ever worked on," says Piatek. That kind of efficiency makes for A-pillars and a windshield header slimmer to see around, and door panels shaped to help create a larger passenger compartment. That passenger compartment also includes additional rearward seat-track movement to better accommodate tall drivers. Accessing the passenger compartment is done via conventional doors rather than scissor doors, which the team felt might be fine for a sunny-day toy but not so great for a car that might also get driven in inclement weather.

A lift-off roof panel has been a Corvette design hallmark going back to 1984, and the C8 team knew they wanted to retain it. The panel is made of polycarbonate or, optionally, carbon fiber, and it weighs approximately 16 pounds. A full convertible, with its own unique design, is also part of the program.

The need to stow the removable roof panel drove the creation of the rear trunk. As Piatek says, "If you take the roof panel off and you have to leave it in your garage and then it rains, it's going to be a bad day."

The rear trunk not only can carry the roof panel, it's also large enough and shaped so two sets of golf bags fit. Meanwhile, the front storage compartment can accommodate a standard airline cabin bag lying on its side. Combined, the total cargo volume is 12.6 cubic feet. "The car has more utility than anything that's comparable in the segment," says Piatek.

TOP: The passenger compartment moves forward a whopping 16.5 inches. Forward visibility is much better, thanks to a much lower cowl and instrument panel. **ABOVE:** A liftoff roof panel has been part of Corvette design since 1984.

ELECTRONIC ARTS

Working behind the scenes in the new Corvette platform in a nonetheless critical role is the car's new digital vehicle platform. Electronics is one of the fastest changing aspects of modern cars, and the C8's new electronic architecture brings with it the capability for over-the-air software updates. Additionally, cyber-security demands encoded messaging between all the modules on board so no one can hack the car. The system also allows for faster signal transmission with BUS traffic four times as fast as before. That is particularly important with active chassis controls, which are dependent on data streams coming from sensors located around the car. The faster that input data comes in, the quicker the car can react to what's going on.

Customers might be more impressed with the new electronic features on the 2020 Corvette. A rear-camera mirror is yet another Corvette first, allowing drivers to switch between a traditional mirror view and a wide-angle image feed from a camera mounted at the rear of the roof. The instrument cluster is a 12-inch fully customizable display, and the standard infotainment system boasts higher resolution and simpler menu logistics. Bluetooth phone pairing also is easier with near-field communication (NFC), and wireless device charging is available. A heated steering wheel is newly available, as is memory for the driver and passenger power seats. A new second-generation performance-data recorder records in higher-definition 1080p and does so with or without an overlay of telemetry data including speed, g forces, gear choice and more. A new feature for the performance-data recorder is it can function as a dash cam, automatically recording every drive and overwriting old files with the new.

ROAD TUNES

Although the C8 team went to great lengths to pipe in the small-block V8's sweet sounds, sometimes you want to listen to different music. That's where Bose comes in. As has been the case since the 1984 C4, Bose provides the tunes in the new Corvette. Because the C8 cabin's size, shape and layout is so radically changed owing to the mid-engine design, the new car's audio system also had to be completely reconfigured. The top-spec Bose Performance Series system features 14 speakers—the most ever used in a Corvette—

driven by a 16-channel digital amplifier. Bose Nd 10-inch woofers are discretely packaged into each door, with the upper door panel home to a 4-inch wide-range speaker and a 1-inch tweeter. At the front of the cabin contained within the instrument panel are 2.5- and 3.25-inch Bose Twiddler speakers, while the rear of the cabin features two 5.25-inch wide-range speakers and three more 3.25-inch Twiddler speakers. Bose AudioPilot technology uses microphones to measure in-cabin sound and allows the system to compensate for ambient noise.

HIDING IN PLAIN SIGHT

Bringing the eighth-gen 'Vette to production required extensive testing, and test mules logged nearly 1,000,000 real-world miles on the road. The keen camera lenses of spy photographers captured images that fanned the flames of mid-engine Corvette fever.

"On a typical Corvette program, we can test right under everybody's nose," Juechter says. Advanced development work on a new chassis, changes to suspension geometry, new tires, new powertrain stuff all can be installed under an old body, which means test engineers can drive around and onlookers would never know it's the new car. This time was different.

Camouflage and altered body panels are often used to obscure a new car's shape, but a mid-engine car's proportions are so different from a front-engine model that they're a dead giveaway. "We talked about putting a giant, fake front end on the thing so it would look like it's got a long hood, but that totally invalidated any of the test work we needed to do," Juechter says. The engineers were able to achieve a fig leaf of disguise by using a Holden Ute body in the early evaluation of the basic chassis and body structure, but testing soon progressed beyond that stage.

"We went to the Nürburgring, we knew pictures would get taken, we knew people would be able to do renderings that were pretty faithful to where the car was going to end up, but we took that as a given," Juechter says. "We had to do representative aerodynamic testing, representative cooling testing. We didn't want to get surprises at the end because we had camo interfering with our primary mission. We resigned ourselves to not be so onerous on our product security that we missed something in testing. And that goes back to the imperative that we get this car right the first time."

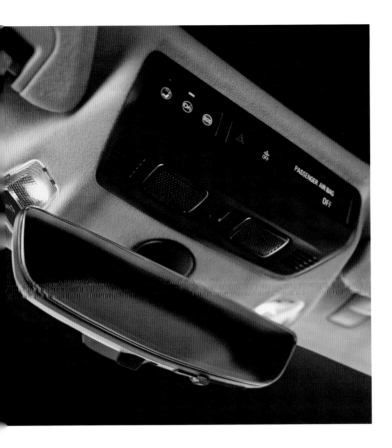

The eighth-generation Corvette logged nearly 1,000,000 real-world miles on the road. Spy photographers' keen camera lenses captured the images that fanned the flames.

NEXT SPREAD: The new Corvette looking good as it poses behind GM Design.

ABOVE: From left: GM President Mark Reuss, design boss Michael Simcoe, CEO Mary Barra, design manager Kirk Bennion, designer Tom Peters and designer John Cafaro check out their new baby.

ONE SHOT

Getting the car right the first time was an imperative—and it was far from a given. "We knew we had our work cut out for us," says Juechter.

As Piatek recalls, "As we started imagining this four or five years ago, we were sitting in a dark room looking at a big screen, and somebody was spinning around the CAD drawings. I really think there was some uncertainty as to how good the car would be in our first try. I mean, there have been other instances where we've done a car that we hadn't done before and it turns out OK but isn't all-conquering out of the gate. And you have to do a couple iterations to really find where some of the sensitivities are and some of the nuances that can make the car really good."

"It would have been easy to basically copy a high-level exotic and try to do what they did," says Juechter. "But we also had other criteria—around affordability, around manufacturability, around serviceability, around luggage volume. There are a whole bunch of other things that we had to do that other people who operate in that space didn't have to do."

And the end result speaks for itself—although chief engineer Juechter also has a few words: "The fact that in our first try, using all the tools we had and having an incredible army of really dedicated people, that the car turned out to be so integrated and so competitive is really ... it's awesome."

AN ATTAINABLE DREAM CAR

If the new car's design could be strongly inferred from the obsessive media speculation leading up to the car's July 18, 2019, reveal and its midmounted engine was already known, then what was perhaps one of the car's most shocking aspects announced at the car's unveiling event was the starting price: less than $60,000.

"We have a lot of customers who can afford anything, but we also know that Corvette owners on average spend a higher portion of their discretionary income than do owners of any other car," says Juechter. "And we don't want to walk away from those people."

That's one reason the car sells so well. It's an aberration in this country to have a sports car like the Corvette sell in the volumes it does. "Of course, we have to make a business case," says Juechter. "We could have decided to price it much higher and taken lower volume. We decided not to go that way. We want this exotic driving performance experience available to as many people as possible."

A NEW BEGINNING

Democratizing performance is what the Corvette has always been about. With its new configuration, the 2020 Corvette Stingray democratizes the exotic-car layout and driving experience as well. And if Reuss' pricing announcement was the most surprising element of his remarks at the C8's launch, his more exciting statement was this: "We are just getting started with Corvette."

TADGE JUECHTER
EXECUTIVE CHIEF ENGINEER, CORVETTE

Growing up in Chappaqua, New York, Tadge Juechter's destiny to become the leader of the team that created a revolutionary new Chevrolet Corvette was hardly preordained—his was a Porsche household. In fact, for Juechter, the very idea of becoming an automotive engineer was hazy at best. Engineer was not a job people had in that well-to-do Manhattan suburb. They were doctors and lawyers and the like. "I didn't have a concept of what an engineer would do as a job," he says. "All I knew is that I loved mechanical things from a young age, and I was always taking stuff apart and making new things out of the pieces."

In high school, he was in honors math and science, but when class was over, he headed down to the shop. "I was more interested in what was going on in the shop than I was anywhere else in the school," he says. "I just thought I was weird. I liked math and science, but I also liked getting my hands dirty, getting my knuckles scraped, learning how to weld or how to run a lathe."

Outside of school, "almost all of my time was spent on inventing stuff." He found a broken lawn mower on the side of the road and put the engine into a go-kart he'd built in shop class. He put engines on bicycles. He built full-suspension bicycles (long before mountain bikes were a thing). "I was always bombing around the neighborhood in some weird contraption."

His parents were supportive. They gave him a 1964 Cadillac Sedan DeVille to dissect after his mother had wrapped it around a tree. "My friends and I disassembled the entire thing, all the way down to taking the transmission apart," he recalls.

In college at the University of Rochester, Juechter ended up in the engineering school, and it was no surprise when he took a job at General Motors in 1977, while still an undergraduate. Juechter worked at GM's Lordstown, Ohio, assembly plant in the summer after his sophomore year and again the following summer. After receiving his Bachelor of Science degree in mechanical and aerospace engineering in 1979, he moved to Michigan to work at GM full time. He earned an MBA from Stanford University in 1986.

Juechter joined GM working at the Fisher Body Division in research and development. His first project was a new door lock. "I worked on lots of little mechanisms like that: hinges, windshield wipers, glass-guidance mechanisms." He also worked in safety research, developing and then crash-testing solutions using Chevy Citations as a test bed, projects done in cooperation with the federal government as it sought to craft new crash regulations to improve safety in a tangible way.

Juechter moved to the Corvette program in 1993. It was during the C4 era, but he was part of the team staffing up to begin work on the fifth-generation car. For the C5 program, he became total vehicle integration engineer. He cycled through various titles during the life of the C5 program, but essentially Juechter became Chief Engineer Dave Hill's deputy. That partnership continued into the C6 program. Juechter effectively became Corvette chief engineer in 2006, but the position was under Vehicle Line Executive Tom Wallace (who also had responsibility for other cars). When Wallace retired in 2008, Juechter's role as Corvette program boss suddenly became more visible.

As chief engineer of the seventh-generation car that debuted for 2014, Juechter's team created an aluminum frame so structurally advanced it could be used under any C7 variant. With the arrival of the wide-body 650-hp Z06 and then the 755-hp ZR1, he had taken the front-engine/rear-drive Corvette to its zenith. And now with the eighth-generation car, he has marshaled a team that has changed the very definition of what a Corvette is and can be.

The dream born with the first Corvette Chief Engineer, Zora Arkus-Duntov, has finally come to fruition some 60 years later, under the leadership of the fifth, Juechter. Looks like those hours he spent modifying bicycles, turning a lathe in shop class and disassembling a crashed Caddy were not wasted.

Of course, Juechter had a great team behind his efforts. Here he talks about some of the key players who helped get the job done.

Chief Engineer Tadge Juechter has worked on Corvettes since 1993. He became chief engineer in 2008.

HARLAN CHARLES
PRODUCT MARKETING MANAGER

"Harlan has been on the program since the C5 days. He was an early advocate for studying mid-engine. Back when Dave Hill was in charge, at a time when people said GM would never produce a mid-engine Corvette, he was gutsy enough to put together one of the original rationales supporting it. He was an out-of-the-box thinker. He is an engineer by trade but worked in design for a long time, so he had his big-picture cap on.

"He and I ended up teaming up ... and we created a technical presentation and marketing-side presentation why this would make sense. And we ended up using that presentation all the way up to the chairman of the board, to get this idea OK'd. Bob Lutz originally was opposed to the idea, but after he saw our rationale, he changed his mind. We eventually (made the presentation) to Rick Wagoner and a lot of other people who have a stake in Corvette's future.

"He's one of the heroes of the car, for sure, and he's been with us all the way through and will be helping lead the communications with customers, as to why what we did was the right thing to do."

ED PIATEK
CHIEF ENGINEER

"Ed Piatek will be new to a lot of Corvette folks. Technically, I'm the executive chief engineer and Ed is my chief engineer. He has integration responsibility for the whole car. He came to us from high-performance vehicle operations; he has a high-performance background.

"He's a very passionate car guy and he's going to be emerging from the shadows. He's been working behind the scenes for almost six years getting this car designed, built and to market."

JOSH HOLDER
PROGRAM ENGINEERING MANAGER

"One of the true heroes on the team. Josh comes to us from Montana. He's a freak because growing up in Montana, his dad was a Corvette guy, and Josh has made it his life's mission to get to work on Corvette. He had to make his way through enormous organizational and other challenges to get into GM and to get to the Corvette team. He took any job he could to get one step closer—always migrating toward Corvette. He finally got to the Corvette team some years ago. His official title is program engineering manager. He's a brilliant engineer. He's super-passionate about the car, he owns a bunch of cars and his wife owns Corvettes, so they are just maniacal about it. Josh works tirelessly and has driven this program through all sorts of hurdles."

KELLY BELLORE
PROGRAM MANAGER

"Corvette is unique in a lot of ways because we have unique interactions where we're face-to-face with a lot of customers; we also have a lot of unconventional activities that go on, and people expect a lot stuff that is outside the mainstream corporate customer responsibilities. The person who coordinates most of that is Kelly Bellore. Kelly has been on this program now for quite a number of years—she effectively is my chief of staff. She orchestrates all the activities, not just technical but everything having to do with program management once you get the car done. So, it's not just engineers. We have people in finance, we have people in marketing, in purchasing, in validation, in testing and development. There are lots of different organizations that have to contribute. Kelly is like the band leader or the orchestra conductor across all those functions."

Mark Reuss, Tadge Juechter, Kelly Bellore, Harlan Charles, Tom Toft and the team have plenty to be proud of as the newest Chevrolet Corvette takes its place on the world stage.

HIGH-PERFORMANCE VARIANTS

Building on the eighth generation's unique design and innovative engineering, Chevrolet is producing ultra-high-performance Corvettes that deliver maximum driving pleasure and the ultimate ownership experience in every scenario.

2023 CORVETTE Z06: THE SUPREME AMERICAN SUPERCAR

Continuing a legacy begun in 1963, the Chevrolet Corvette Z06 is designed and engineered to be a precision tool that's equally at home on the road or on the track.

When the Z06 option code was first introduced in 1963, it immediately established a new standard for performance among production sports cars. The Z06 "Special Performance Equipment" package included race-tuned suspension, steering and brakes, as well as a potent fuel-injected V8 engine coupled to a four-speed transmission and limited-slip differential. Combined with aluminum knock-off wheels and a 36-gallon fuel tank, it transformed standard Corvettes into formidable road racers. Remarkably, though, these cars were equally suited for long-distance touring and everyday driving.

In 2001 Chevrolet revived the Z06 designation, turning it into a model rather than just an option package. The C5 Z06 lived up to its forebearer's legendary status among enthusiasts, courtesy of a stiffer structure, added power, more aggressive gearing, revised suspension, stickier tires and lower mass; together, these innovations yielded the fastest, best-stopping, best-handling Corvette ever produced to that point. C6 and C7 Z06s followed, each raising performance to a new high with cars that pointed the way to the C8 Z06, which would set remarkable new benchmarks both on and off the track.

"We designed the eighth generation to do a really big bandwidth of cars from the very beginning," explains Corvette Product Marketing Manager Harlan Charles, "and we saw the Z06 as a centerpiece that would establish new levels of performance. We looked at every aspect of the car in order to bring the most up-to-date racing technology to the street, but true to its heritage, this is a car that is attainable and fun to drive in all circumstances."

THE HEART OF THE BEAST

Since its debut all the way back in 1955, Chevrolet's V8 engine has led the industry with an unmatched combination of high power, low mass, durability, efficiency and simplicity. After nearly seventy years of development, however, engineers were bumping up

TOP: The Z06 performance package, which included a 360 horsepower fuel-injected engine and heavy duty suspension and brakes, turned 1963 Corvettes into successful road racers.

ABOVE: A long list of special features, including its 650 horsepower supercharged LT4 engine, select carbon fiber body panels, augmented cooling, and bespoke suspension tuning made the C7 Z06, introduced in 2015, the highest performing Corvette to that date.

2023 Corvette Z06 is available as a coupe or a convertible. Because the underlying body structure is so strong, no additional bracing is needed for the convertible.

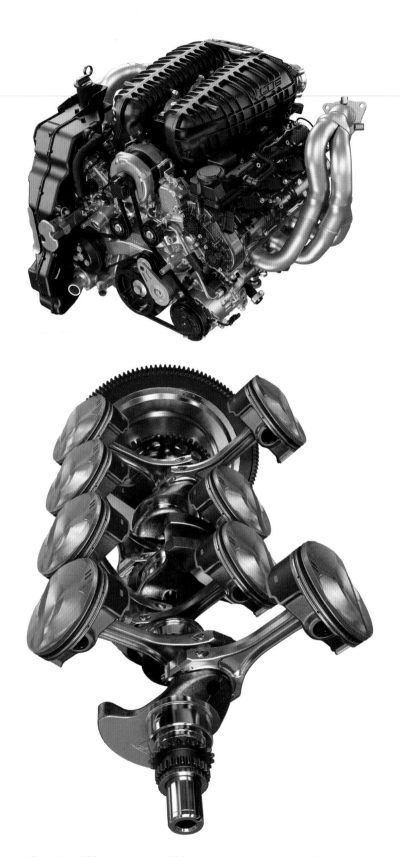

against the limits of physics and realized that a car as revolutionary as the C8 Z06 would need an equally advanced powerplant. The result of their efforts, a 5.5-liter engine called LT6, produces 670 hp at 8,400 rpm and 460 lb-ft (624 N-m) of torque at 6,300 rpm, making it the most powerful naturally aspirated V8 ever offered for a production car.

Testifying to just how advanced this new engine is, GM's propulsion engineers code named it Gemini, after the early United States space flight program, and just for fun incorporated no fewer than fifty-four decorative rockets in its crankcase, cylinder heads and various other castings. The cylinder case and oil sump are made from aluminum and mated at the crankshaft centerline. In a stark departure from all previous GM engines, the LT6 uses a race-inspired 180-degree flat-plane crankshaft. This type of crankshaft is lighter than a conventional cross-plane crank, and thus spins up quicker and revs higher, contributing to the LT6's 8,600-rpm redline. In addition, with a flat-plane crank the exhaust pulses in each cylinder head are spaced 180 degrees apart for more efficient breathing.

Lightweight and extremely strong forged-aluminum pistons supplied by CP Carrillo feature domed heads that yield 12.5:1 compression. The pistons, which ride in cast-iron cylinder liners, are anchored to Pankl Racing Systems connecting rods forged from titanium for the best combination of high strength and low mass. The engine's short 3.250-inch stroke, in concert with its generous 4.104-inch bore, yields 5.5 liters of displacement. It's no coincidence this is the maximum displacement allowed for production class-racing in IMSA and FIA World Endurance Championship competition, where an all-out race version of the LT6 powered Corvette C8.Rs to victories at Sebring, Daytona and Le Mans as well as IMSA and WEC championships since its debut in 2020.

Adding to the innovations found in the new engine, a hollow, crankshaft-driven intermediate shaft between the cylinder banks turns two plunger-type 5,000-psi fuel pumps as well as four overhead camshafts. The camshafts, which are also hollow to minimize mass, actuate two 1.654-inch (4.201-centimeter) titanium intake valves and two 1.378-inch (3.500-centimeter) sodium-filled, stainless steel exhaust valves per cylinder for excellent air flow. Breathing is further enhanced by CNC-machined intake ports and

TOP: The all-new LT6 powering every Z06 is the most advanced engine GM has ever manufactured and, with 670 horsepower at 8,400 rpm and 460 lb-ft (624 N-m) of torque at 6,300 rpm, the most powerful naturally aspirated V8 ever offered for a production car.

ABOVE: The LT6 engine's flat-plane crankshaft is lighter and more efficient than the cross-plane design found in all prior Corvettes. Combined with forged aluminum pistons and forged titanium connecting rods, it delivers unprecedented performance in the C8 Z06.

combustion chambers in each cast-aluminum cylinder head. Adjustable, computer-controlled camshaft drive sprockets allow for 55 degrees of timing variability for the intake valves and 25 degrees for the exhaust valves, one more measure that allows the Z06 to be both a ferocious track car and a civilized street car.

Twin 87mm throttle bodies meter induction air to large molded-nylon plenum chambers that each have a volume equal to the engine's 5.5-liter displacement. Three computer-controlled valves between the two plenums open and close to take full advantage of the natural pressure waves induced by the action of the intake valves. By controlling these pressure waves, the induction system takes advantage of a phenomenon called Helmholtz resonance to magnify the amplitude of the air in the plenums, effectively forcing the intake charge into the cylinders and allowing the engine to exceed 100-percent volumetric efficiency. The real-world benefit of this is a torque curve that remains magically close to a peak output of 460 lb-ft (624 N-m) all the way from 3,500 rpm to the LT6's 8,600 rpm redline.

The LT6's exhaust system also employs advanced technology to enhance overall performance. Upward-sweeping Y-shaped headers crafted from stainless steel are precisely configured to minimize back pressure and maximize the scavenging effect induced by the flat-plane crank's 180-degree exhaust pulsing. The advanced engineering in the exhaust system continues all the way back to the four center-outlet tips, which were designed with a "reverse megaphone" shape to reflect sound off the parabolic surfaces on the car's rear body panels and redirect it toward the passenger cabin. The result is an aggressive yet melodious harmonization of combustion notes that enhances the overall driving experience.

A RACE CAR FOR THE STREET

The Z06 retains the base Corvette's Tremec TR-9080 eight-speed, dual-clutch transmission, but it is fitted with a lower, 5.56:1 final drive ratio for quicker acceleration. The electronically controlled limited-slip differential optional for Stingray is standard with Z06. GM's fourth-generation Magnetic Ride Control (MRC) suspension, introduced as optional for the Stingray in 2020, is also standard in this model. MRC 4.0 benefits immensely from the use of

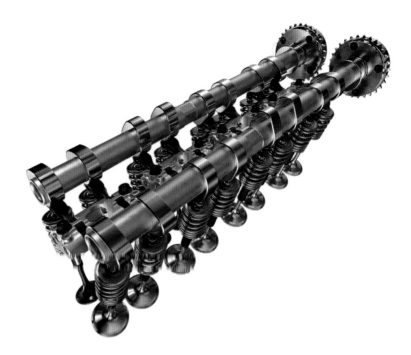

accelerometers on all four corners of the car rather than position sensors to indicate heave, roll and pitch. Precision and tunability of the system are also improved by the incorporation of an inertial motion unit that measures actual rather than calculated movement signals. For Z07-equipped Z06s, MRC gets a unique calibration that enhances track performance.

To take full advantage of the amazing MRC 4.0 suspension, Z06 is fitted with more aggressive bodywork that measures 79.7 inches (202 centimeters) wide, which is 3.6 inches (9.1 centimeters)

TOP TO BOTTOM: An incredible amount of engineering went into designing the center-exit exhaust, available only with US-delivered Z06s. The tips use a "reverse megaphone" shape to reflect sound off the parabolic surfaces on the car's rear body panels and redirect it toward the passenger cabin.

wider than the Stingray. This provides the necessary clearance for 20 × 10-inch (508x254-millimeter) front and 21 × 13-inch (533 × 330-millimeter) rear forged-aluminum "spider" wheels, which are available in five different finishes. Michelin Pilot Sport 4S ZP tires, sized at 275/30ZR20 and 345/25ZR21 respectively, are standard fare while the optional Z07 package includes even stickier Michelin Sport Cup 2 R ZP tires that were developed specifically for Z06. In either case, the front tread is 30mm wider than Stingray tires and the rear tread is 40mm wider, contributing to the car's enhanced handling and more aggressive appearance.

For added stopping performance, Z06 gets a bespoke Brembo brake package, with larger 14.6-inch-diameter (370 millimeter) front and 15-inch-diameter (380 millimeter) rear rotors and six-piston front calipers, compared to four-piston calipers on Stingray. Brakes are further enhanced with the Z07 option, which employs vented, cross-drilled carbon-ceramic rotors and compatible pads. The fronts measure 15.7 × 1.5 inches (399 × 38 millimeters) and the rears are 15.4x1.3 inches (391 × 33 millimeters).

To contend with the added heat a Z06 can generate, on-track engineers specified several important enhancements. The unique front fascia incorporates larger air intakes and, like the C8.R race car, draws clean air into a central heat exchanger, one of five located throughout the car to provide additional cooling air to the engine, brakes and transaxle. The wider rear fenders also allow for larger air intakes than those found on the Stingray.

Besides adding cooling capacity, Z06's bespoke bodywork also improves aerodynamic performance. And, as with the C7 Z06, multiple levels of aerodynamic trim are available with the new Z06. The base setup features a front splitter and rear spoiler with a removable wickerbill that generates up to 365 pounds of downforce at 186 mph.

Aero performance can be enhanced with optional Carbon Fiber (RPO CFZ) and Visible Carbon Fiber (RPO CFV) Performance packages. Corvettes equipped with CFZ or CFV can also get an optional Carbon Aero Package with large rear spoiler and front dive

LEFT, TOP TO BOTTOM: The Z06 body measures 3.6 inches (9 centimeters) wider than a Stingray. The added width accommodates wider tires, improves aerodynamic performance, and delivers a far more aggressive stance; Optional carbon-fiber wheels eliminate slightly more than 10 pounds (4.5 kilograms) of unsprung mass and rotational inertia from each corner. A carbon ceramic brake system included with the Z07 package eliminates more mass and delivers incredible stopping power; In keeping with the Corvette Z06 tradition that dates all the way back to 1963, the C8 Z06 is equally at home on a racetrack or the street.

planes, which increases rear downforce and dramatically changes the look of the car. Called TOF/TOG, Carbon Aero Package is required with the Z07 performance package.

The highest-performing aero package comes with option Z07, which includes a larger front splitter, front-corner dive planes, a rear wing and underbody strakes. When adjusted for maximum effect, the Z07 aero components deliver up to 734 pounds (333 kilograms) of downforce at 186 mph (299 kph).

For even higher levels of performance, carbon fiber wheels are optional with the Z07 package. Compared with the standard forged aluminum wheels, the Carbon Revolution–supplied one-piece composite wheels save about 41 pounds (19 kilograms) of unsprung mass. A Z06 riding on lightweight carbon wheels and equipped with a Z07 package—which includes the precisely balanced, high-downforce aero package, specifically calibrated FE7 suspension, carbon-ceramic brakes and sticky Michelin Sport Cup 2 R ZP tires—is a fearsome track warrior indeed.

Though ultra-high-performance is clearly the central focus for Z06, it doesn't come at the expense of comfort and luxury. The Stingray's already fabulous interior is further upgraded for Z06 with two optional carbon-fiber trim packages and a Stealth Aluminum trim package. Combine these with the many other cosmetic choices for both the interior and exterior and Z06 buyers can truly build the car of their dreams.

2024 CORVETTE E-RAY: A SUPERCAR FOR ALL SEASONS

The E-Ray's creators call the electrified Corvette their "Swiss Army knife," because it has multiple features that enable it to do a lot of things extraordinarily well. "While the Z06 is like a scalpel, a razor-sharp tool for the track that's equally at home on the street," explains Harlan Charles, the E-Ray is more like a Swiss Army knife

ABOVE, TOP TO BOTTOM: Every inch of every part on the Z06 was designed to maximize performance. Adding the optional Z07 package to a Z06 ups the ante, dramatically transforming both its performance and appearance, courtesy of a larger front splitter, front-corner dive planes, a large rear wing and underbody strakes. Each of these components is crafted from carbon fiber to minimize mass; the quality of workmanship and materials used in all C8 interiors is remarkable. In addition to adding any option available for the Stingray interior, Z06 buyers can also specify one of two optional carbon-fiber trim packages and a Stealth Aluminum trim package.

ble driving experience, but they do it in very different ways."

The E-Ray is the most technologically sophisticated and diversely capable Corvette ever produced. Its conventional internal combustion engine, combined with a potent electric motor, give it the total power output of a supercar, and its all-wheel drive puts every bit of that propulsive force to the ground. The result is the fastest-accelerating Corvette in the marque's history, a car that can be driven safely and effectively on less-than-ideal road surfaces and in even the most severe weather conditions.

OPEN SPACE

Though the move to mid-engine architecture dictated that the C8 be completely redesigned, one of several important elements from prior-generation Corvettes that did carry over was a central structural tunnel. Because the tunnel, initially introduced for C5, is no longer needed to house a driveshaft, its 2.5-cubic feet of space is available for other things.

"The mid-engine layout means there's not a lot in that tunnel," explains Harlan Charles,

so it provides an opportunity to do new things without making the car bigger or taking away from its storage space. Customers have been asking for all-wheel drive, both for performance and for more usability throughout the year in places where it rains and snows. From the beginning, very early in the development of this architecture, we talked about implementing a more conventional all-wheel drive, with a drive shaft going through the tunnel to a front differential. While that was certainly feasible, we quickly concluded that adding electrification to arrive at all-wheel drive was the best solution for several reasons, including the fact that it adds more power and torque.

Though there were numerous technical hurdles to overcome—most notably the complete integration of the electric-powered front wheels with the internal-combustion-powered rear wheels to create what Chevrolet calls eAWD—this fusion of propulsion sources delivers the best that both have to offer. E-Ray drivers have the freedom and convenience that conventional fuel provides

TOP: Whether as a convertible or a coupe, E-Ray gets the same wide bodywork used for Z06, giving it an equally beautiful stance.

with the instant torque and near-silent operation that an electric motor gives, married in a package that yields unprecedented levels of traction in all conditions.

MAKING POWER

The E-Ray's internal combustion engine is the 495-hp LT2 V8 found in the Stingray equipped with optional performance exhaust or Z51, driving the rear wheels through the same highly capable Tremec TR-9080 eight-speed, dual-clutch transmission used for Stingray and Z06. The only change to the automatic transmission is the addition of an electric hydraulic pump to enable clutch operation when the LT2 isn't running and the car is driven purely with electricity. When the engine is running, the electric pump remains idle and clutch operation takes place via the standard engine-driven hydraulic pump.

The electrified Corvette's front wheels are driven by a permanent-magnet AC motor mounted beneath the front storage compartment. The motor, produced by GM Subsystem Manufacturing LLC in its Brownstown, Michigan, assembly plant, is only 7 inches (18 centimeters) in diameter but produces an impressive 160 hp and 125 lb-ft (169 N-m) of torque. The motor's remarkable size-to-power ratio, in conjunction with clever packaging, enabled Corvette's engineers to reach their performance targets without adversely impacting the car's storage capacity.

The compact motor's twist is transferred to the front wheels by means of an open differential fitted with an 8.2:1 ratio reduction gearset inside a magnesium housing. To provide velvety-smooth operation, help with cooling and maximize longevity without accompanying parasitic power loss, the front differential is lubricated with its own dry-sump oil system.

E-Ray's electric motor is powered by an amalgam of eighty pouch-type lithium-ion battery cells manufactured by LG Energy. Total capacity for the battery pack is 1.9 kWh, but to reduce system stress and increase the battery's longevity, only 1.1 kWh is used in most driving situations. The battery's chemistry and accompanying software and hardware are configured to maximize

RIGHT, TOP TO BOTTOM: E-Ray's interior is distinguished by a special badge at the bottom of the steering wheel, as well as electric output and battery information on the screens. As with Stingray and Z06, E-Ray buyers can choose from among eight interior trim colors and three seat styles; The E-Ray is a tour de force of high technology that is equally adept at grand touring, track days and everyday commuting; To help offset the added weight from E-Ray's motor and other electrification components, the car comes standard with carbon ceramic brakes. The five-spoke aluminum wheel design is unique to E-Ray.

rapid power delivery rather than extended range, as is typically found with GM's hybrid vehicles.

There is no provision to plug the E-Ray's battery pack into an external electric supply, so it must be charged from a combination of three onboard sources. During coasting and braking, the car's AC motor goes into regenerative mode and acts as a generator, using the kinetic energy of the car's motion and braking action to charge the battery. A third charging mode, called Charge+, effectively uses the LT2 engine to turn the AC motor/generator, enabling it to recharge the battery when cruising and accelerating. The Charge+ mode, which is selectable by the driver, combines with regenerative charging from braking and lift-throttle coasting

ABOVE, TOP TO BOTTOM: Beautifully styled badges on the rear quarter panels call out the electrified Corvette's name, which was dreamed up by Product Marketing Manager Harlan Charles.

to rapidly replenish the battery to full power. With coasting and braking regenerative charging only, a completely depleted E-Ray battery will be fully recharged after approximately 3 miles (4.8 kilometers), but with Charge+ this is reduced to as little as about 1 mile (1.6 kilometers).

MINIMIZING MASS

As the automobile industry transitions to electric power, one of the biggest challenges facing engineers is managing the added mass that goes along with designing for electrification. The weight of the battery pack, motor, inverter, cooling loops, wiring and all related components quickly adds up. An additional structural chassis support, necessitated by the upward relocation of front coil-over dampers to make room for axle shafts connecting the differential's twist to the front wheels, also added weight to E-Ray. The law of physics dictates that added mass directly reduces performance in virtually every measure, making it imperative to minimize the effects of mass wherever and whenever feasible.

To address this issue, engineers began with a relatively small, compact yet powerful system. "In a general way, we patterned it after the KERS system used in Formula One," reveals Harlan Charles. "It's modestly sized to save weight and allow for very efficient packaging." Beyond that, Chevrolet's engineers looked at each individual component in the electrification system to develop clever design innovations and material specifications that reduce mass. For example, the front differential's housing, cast in lightweight magnesium, is further lightened using aluminum rather than traditional steel fasteners. The car's battery pack housing is fabricated with composite plastic, further minimizing weight, and the system's three cooling loops were laid out to be as mass efficient as possible.

Beyond electrification-specific components, the engineers took a whole-car approach to minimizing E-Ray's mass. The Brembo carbon-ceramic brake system, optional on Z06, is included with E-Ray, making 41 pounds (19 kilograms) of unsprung weight disappear, and Corvette's standard 12-volt battery that turns over the

LT2 engine and helps power the whole car gives way to a lighter lithium-ion design. All these mass-saving moves together keep E-Ray's overall weight at just 275 pounds (125 kilograms) more than the Z06.

SOPHISTICATED INTEGRATION

A highly unusual characteristic of E-Ray is the absence of any physical connection between the front and rear wheels. For both safety and performance, it's imperative that the action of each wheel in an all-wheel drive system be coordinated with the other wheels. Without a physical link connecting the front and rear, all of this integration has to be done electronically.

Mike Kutchner led the engineering team that created E-Ray's leading-edge eAWD integration system. The design includes sensors at each wheel and elsewhere in the car to accurately measure what's occurring, along with highly sophisticated computers running advanced software to adjust power input and braking action; working in tandem, these components help keep the car stable while maximizing performance in all conditions. For the latter, that includes both ends of the performance spectrum: achieving

the fastest acceleration possible when called for in all weather and road conditions, and optimizing fuel efficiency when the Active Fuel Management system is engaged.

Working closely with Cody Buckley and the other Corvette development drivers, Kutchner's team continuously fine-tuned E-Ray's control strategies to achieve utterly seamless integration between the gasoline-driven rear and electricity-driven front wheels. The real-world result is the imperceptible addition and removal of supplemental power to the front wheels as needed to make the electrified Corvette blazingly fast, sure-footed in every season on every road surface and intensely fun to drive.

PUTTING THE POWER DOWN

E-Ray delivers on its fundamental mandate to deliver user-friendly, supercar performance year-round, regardless of weather, with new all-season tires. Developed with longtime partner Michelin, the rubber is sized at 275/30ZR-20 up front and 345/25ZR-21 in the rear, making these the widest all-season tires produced. They ride on five-spoke aluminum wheels designed specifically for the E-Ray. If the tire size sounds familiar, that's because these are the

ABOVE: The E-Ray's eAWD system delivers a noticeably different driving experience than Stingray or Z06 because of the added traction its electric motor-driven front wheels provide.

same specifications used for super-sticky Z06 Pilot Sport Cup tires. To house the massive all-season rubber, E-Ray gets the same bodywork as Z06, and thus also measures 3.6 inches (9 centimeters) wider than a standard Stingray. To help differentiate E-Ray from Z06, however, the former has body color trim while the latter is adorned with carbon flash black trim accents.

As with the Stingray, E-Ray drivers can choose from among six driving modes—Tour, Sport, Track, Weather, the customizable MyMode and Z Mode—allowing personalization of almost every aspect of the driving experience. The distinctive driving characteristics that E-Ray delivers is felt in each mode, and the more one experiences it, the more this groundbreaking car's unique personality shines through. "It offers incredible acceleration," Harlan Charles points out, "and we tied it in with motorsports technology with the Charge+ button, which allows you to recharge the system very quickly and then spend all that energy for maximum acceleration once again. Beyond that, it's a great canyon carver, and it has all-weather capability. Plus, you have the stealth mode and can run it as a pure EV for a few miles for the quiet exit of your neighborhood, so it's really a new kind of Corvette. It offers

performance approaching a race car, but in a more sophisticated and comfortable way."

As far as race-car-like performance is concerned, E-Ray's numbers tell the story. The car's 655 combined horsepower and eAWD propel it from 0–60 miles per hour (0–97 kilometers per hour) in 2.5 seconds, making it the fastest-accelerating production Corvette ever made, and one of the fastest street legal cars in the world, at any price point. Equally impressive, E-Ray will erase the quarter mile from a standing start in a mere 10.5 seconds at about 130 miles per hour (209 kilometers per hour).

THE LOGICAL SUPERCAR

While E-Ray and Z06 performance numbers are strikingly similar, each arrives at the extremes of its performance envelope in a different way. The differences are what will determine which car is the best fit for a given buyer. "It's a different hierarchy for Corvette," Harlan Charles explains.

And I like to analogize the differences to *Star Trek* characters. The Z06 is similar to Captain Kirk. He's very emotional, powerful and physical, and that fits the Z06's personality. Then there's Spock. He is extremely logical and scientific. He's still very strong, but he approaches problems differently, in a more sophisticated way. That's the E-Ray. The Z06 gets to its impressive performance more like Kirk and the E-Ray gets there more like Spock. They're both heroes in their own way and they both get things done but they approach life from different directions. That illustrates the two sides of our Corvette variants because they have different personalities and they achieve very similar results in performance but in different ways.

CAUSE FOR CELEBRATION: THE 70th ANNIVERSARY EDITION

Corvette is the oldest continuously produced car model in history and Chevrolet has accommodated the wishes of its loyal fan base with special anniversary models going all the way back to the first milestone—the twenty-fifth anniversary in 1978. To honor the nameplate's seventieth anniversary in 2023, Chevy offered a uniquely styled 70th Anniversary Edition.

Unlike some previous anniversary offerings, the 70th Anniversary Edition Corvette can be had with any of the available models: Stingray or Z06, in coupe or convertible. The starting point for any of the models are the 3LT or 3LZ top trim levels, and the most striking feature is clearly the exclusive body paint color, either White Pearl Metallic or Carbon Flash Metallic.

"The idea of offering White Pearl Metallic actually goes back to the sixtieth anniversary in 2013," says Harlan Charles, "but the paint shop wasn't quite ready for it back then. We updated the paint shop in 2018, which has the technology to do that now, so we were very excited to offer it to our customers."

Carbon Flash Metallic also has some history that stretches back about a decade, when it was offered for the 2012 Centennial Edition, but it hasn't been available since. "We looked at different blacks," explains Harlan, "but kept coming back to the Carbon Flash Metallic because it has a lot of fans, and really stands out, just like the tri-coat white."

Corvette Exterior Designer Manager Kirk Bennion developed an optional full-length body stripe package for both anniversary colors. He was determined to make the stripes complementary in a subtle way and succeeded with Satin Grey over White Pearl Metallic and Satin Black on the Carbon Flash Metallic. Rather than contrasting sharply, as stripes typically do, these combinations are more tonal in nature, so they are clearly visible and unify the body front-to-rear, but they don't jump out as a central feature.

The exterior of each 70th Anniversary Edition is further distinguished by visible carbon-fiber mirrors, special badging and bespoke wheels. The multispoke-design wheels are dark grey, with an anniversary center cap and a rather bold red accent stripe around the outer perimeter. The red stripe nicely matches Edge Red brake calipers.

The red wheel accents and red calipers coordinate beautifully with the special 70th Anniversary Edition interior. Brett Golliff, Chevrolet's Global Color and Trim Design Manager, created a two-tone Ceramic White/Black with red stitching interior motif for C8 early on; although the combination didn't make it to the regular interior color palette, it was saved for the special edition. Besides looking stunning on its own, the white-and-red combination is perfect because it evokes the 1953 Corvette's color combination of Polo White exterior and Sportsman Red interior. The theme is further strengthened by the addition of red seat belts.

Anniversary editions came with either GT2 or Competition Sport seats with suede microfiber inserts and 70th Anniversary logos. And, just for fun, the two-tone Ceramic White and Black seat motif is flipped, depending on which seat is chosen: white in the middle with black perimeter accents for the GT2 seats and vice versa for the Competition Sport seats.

Another feature exclusive to the 70th Anniversary Edition is a microfiber-wrapped steering wheel with the 70th Anniversary logo. The same logo is found in the sill plates and in the custom luggage set that matches interior color and fits precisely in C8's available storage space. Finally, the significance of the 70th Anniversary is emphasized with two features found in all 2023 Corvettes, not only the 70th Anniversary Editions: the anniversary logo is contained on a plaque in the center speaker between the seats and etched in the rear window.

TOP: To honor Corvette's milestone 70th anniversary in 2023, Chevrolet offered anniversary editions in either White Pearl Metallic or Carbon Flash Metallic. **INSET LEFT TO RIGHT:** The ZZ3 Engine Appearance Package adds carbon fiber trim and engine lighting to the Z06's LT6 engine. Seventy years after the first one rolled off its assembly line in Flint, Michigan, Corvette is the oldest continuously produced car model in history; All 70th Anniversary Edition Corvettes get these unique multispoke wheels with special center caps and a red stripe around the perimeter that matches the red brake calipers and red interior accents.

ACKNOWLEDGMENTS

The publisher wishes to thank: GM's senior leadership team for its commitment to making Corvette's mid-engine destiny a reality, especially Mary Barra, Mark Reuss and Ken Morris; the team that engineered and developed the first mid-engine Corvette, led by Tadge Juechter and Ed Piatek, as well as the design team, led by Kirk Bennion, John Cafaro, Tom Peters and Phil Zak; and those who were interviewed for this book, including: Kirk Bennion, Harlan Charles, Christo Datini, Brett Golliff, Tadge Juechter, Vladimir Kapitonov, Mike Kociba, Jordan Lee, Alex MacDonald, Tristan Murphy, Steve Padilla, Ed Piatek, Michael Petrucci, Rich Scheer, Mike Simcoe, Ryan Vaughan and Phil Zak. Special thanks to Brenda Eitelman from the GM Licensing Team and Larry Kinsel from the GM historical services and media archive team, as well as Chris Bonelli, Kelly Cusinato, Joe Jacuzzi, Kevin Kelly and Kelly Wysocki from the Chevrolet Communications Team for arranging interviews and helping with assets to make this book possible. I wish to offer a special thank you to Harlan Charles, Corvette Product Marketing Manager, for all of his time and assistance in helping me write the second edition of this book.